MANUFACTURING
Implementation Strategies
That Work

A Roadmap to Quick and Lasting Success

MANUFACTURING

Implementation Strategies
That Work

John W. Davis

Industrial Press
New York

Library of Congress Cataloging-in-Publication Data
>
> Davis, John W., 1938-
>   Lean manufacturing implementation, strategies that work : a roadmap to
> quick and lasting success / by John W. Davis.
>   p. cm.
>   Includes index.
>   ISBN 978-0-8311-3385-6
>   1. Manufacturing processes--Waste minimization--United States. 2. Lean
> manufacturing--United States. I. Title.
>   TS169.D394 2009
>   658.5--dc22
>   2009003156

Industrial Press, Inc.
989 Avenue of the Americas
New York, NY 10018

Sponsoring Editor: John Carleo
Interior Text and Cover Design: Janet Romano
Developmental Editor: Robert Weinstein

DEDICATION

This book is dedicated to the many past associates I came to know at United Technologies, and with special regard to Carrier Air Conditioning, where I spent twenty years of my career before retiring in 1998 and starting a consulting business. During my tenure, I was fortunate to have the opportunity to serve as plant manager for a facility that went on to become a showcase and the catalyst for UTC's "Flexible Manufacturing" process.

To each and every member of that team, including the leadership of the local union, I would like to express my deepest gratitude. Working tirelessly over a three year period, under some extremely challenging conditions, they went on to make some truly remarkable accomplishments.

Table of Contents

FOREWORD

As someone who "grew up" in manufacturing before assuming the role of plant manager at the age of 49, I found that any confidence I had gained in achieving that goal was quickly challenged when the company announced its intention to close three of five manufacturing facilities in the United States and combine manufacturing operations into two sites.

It posed an interesting dilemma for a first-time plant manager who had been in the job less than three months. But it quickly became apparent that the key to insuring the future of an existing facility was to clearly demonstrate the ability to take on additional product, with the least amount of physical expansion required.

With the help of some very good people, I set out to find the best way of doing this and was immediately attracted to a process just taking root in American industry that was built on the principles of the Toyota Production System. Being as steeped as anyone in a batch production mindset, it wasn't easy for me to accept the kind of change being suggested. But intent on learning more, I invited the leadership of the local union to accompany me and my staff to a seminar conducted by Richard J. Schonberger, the author of *Building A Chain of Customers,* and the man who coined the phrase "World-Class Manufacturing."

Realizing the difficult challenge our plant was facing, we collectively came away with a determination to learn all we could about the process and, in the shortest time possible, make the kind of change needed to free up space and improve operating expenses. I can't stress strong enough, however, the critical importance of having the leadership of the local union with us, and of gaining their support as we went about pursuing a new and totally different system of production.

Over the course of the next 18 months, we were able to make a startling turnaround of a 500,000-square-foot Carrier Air Conditioning facility, located in the Midwest. Over 150,000 square feet of manufacturing space was cleared, making room to downsize and close an off-site finished goods warehouse, along with providing the ability to take on a series of products from another factory, without the expense of brick and mortar. With the help of the union leadership, we were also able to rid the plant of a long-standing piece-work wage incentive plan and install a 4X10 (four day, ten hour) work schedule, which required changing an overtime provision in the labor contract from anything over eight hours a day, to anything over forty hours a week.

All in all, it was a truly phenomenal accomplishment, which ended up setting the framework for the plant to become a showcase facility for United Technologies Corporation and the catalyst for UTC's "Flexible Manufacturing" process. As a result, I was later asked to give up my role as plant manager to work in the development and deployment of the process, corporate-wide.

It was indeed a difficult change to make because I was giving up a job as plant manager over one of the company's largest manufacturing operations to essentially take on the role again of an individual contributor. On the other hand, it was also a great honor that the corporation had decided to use our factory as the flagship and principle example for the job to be done.

Due to the extensive travel involved, the "tour of duty" was initially set for two years. My team, who I would not feel I've done justice to without mentioning personally, consisted of Barry O'Nell and Ed Cannon from Pratt and Whitney, along with Gary Rascoe from Otis Elevator. We went on to put in a second tour, which over the course of a four-year period took us to six different countries, interfacing with manufacturing facilities in the United States, Europe, South America, and the Far East, that produced products ranging from fractional H.P. motors and air conditioning to elevators, jet engines, and helicopters, among others.

Those who participated in this near exhausting, but highly rewarding experience gained the equivalent of a Master's Degree in Lean Manufacturing. There was little we didn't see, ranging from extremely enthusiastic plant managers and participants to those who were totally resistant to change. In every case, however, the power of the process won out and the results more than spoke for themselves. One of the more impressive feats was that an entire Otis Elevator facility, located in Mexico, was completely re-arranged, substantially improving productivity and reducing manufacturing lead time from an average of 2 1/2 weeks, to slightly less than 3 days — an astonishing improvement for a two-week period of time!

It served to prove the power of both a well-focused commitment and what can take place over a short period when a cross-functional group, involving high level managers and key factory participants, work together with a clear mission in mind. The effort, of course, was aimed at building consensus, support, and confidence in the process, which under the unyielding commitment of George David, the President and CEO, was very effectively achieved.

I learned many things from the experience. One of the more important was that factory managers and production employees are no different with respect to attitude, commitment, and work ethic, whether they're located in Europe, the Far East, the United States, or elsewhere. Therefore, what we're essentially dealing with is a level playing field, from a production operator standpoint, which makes applied implementation strategy all the more important in the competitive race for the future.

Much of what will be addressed in this work will hit at the very core of conventional thinking and could appear critical of existing efforts. That is not the intention. In fact, the considerable efforts that have extended by many in striving to make Lean a success should be strongly applauded. The flaws and misconceptions that will be pointed out haven't happened because there was any lack of dedication. They have come due to a general misunderstanding as to where the initial thrust of the process should be directed.

The chief enemy we face is time. Given enough time we can probably gain parity with the likes of Toyota and others. But time isn't on our side. Factories across the United States are being downsized and closed at an alarming pace, impacting the welfare of communities and often throwing life-long employees out of work and on their own, to face the challenge of building new and often far less lucrative careers for the future.

As someone who went on to author two leading books on the subject — *Fast Track to Waste-Free Manufacturing* and *Leading the Lean Initiative* (published by Productivity Press) — I have an overwhelming respect for the process. However, it did not fully occur to me until I was actively consulting that finding plant managers who approached the task with the same level of zeal, commitment, and sense of urgency was far from common. This isn't to say they were any less driven to make their operation a success. But they simply were not as fully convinced as I was, at the time, to the unquestionable importance of inserting Lean Manufacturing to its fullest in their respective facilities.

In *"Fast Track"* I pointed out the job to be done was a concerted focus on *completely destroying the existing system of production and replacing it with one that effectively eliminated the extensive wastes and inefficiencies associated with batch manufacturing.* Although some very earnest efforts have been applied throughout industry in America, the strategic emphasis for Lean has not been placed on the *absolute elimination of batch manufac-*

turing, in any shape, form or fashion. The principle reason is two very distinct mindsets that tend to permeate American industry:

Mindset #1 A belief that Lean is just another incremental improvement process like SPC, Quality Circles, and others in the past. No worse, no better. Those with this mindset work at utilizing Lean in spots and continue to perform the business of production in essentially the same manner. But by providing cursory support, if and when they're asked if they are "using Lean," they can always give an agreeing nod.

Mindset #2 A far more common conviction that the only viable method of implementing Lean is to follow the typical approach, which goes something like this: Communicate the need, train a set of cross-functional employees, establish a pilot area to showcase how the process should look, and, from there, spread needed training across the workforce, with as an intense level of "Kaizen" as can be justified.

While we can't entirely dismiss the first perception noted and should do all we can to change it, the principle focus has to be on those who are earnestly striving to make Lean a success. But to a large extent, their efforts have been minimized, due to a growing trend toward cost justification, with a sharp focus on payback on investment. This has come because creeping implementation has provided little actual change to the existing system of production and initial expectations have only vaguely been achieved. This, in turn, has opened the door to other company initiatives, which tend to gain momentum and put a drain on applied resources.

With this in mind, I set out in the spring of 2007 to find the most effective way to go about the task. In doing so, I came to find some relatively serious flaws and misconceptions, most of which center on what was first learned about the process, from a seasoned group of ex-Toyota managers who came to America to teach the basics of the Toyota Production System.

The group was employed by a number of the larger and better known corporations in the United States. They spent time working with key managers and other participants, teaching the importance of the tools and how to go about using them. But in doing so, the tools were never ranked in order of importance. Kaizen (continuous improvement activity) ended up taking the lead role in implementation strategy — to a large extent at the expense of some far more important elements of the process.

What is important to understand is that the tools and techniques cannot be looked upon as a smorgasbord of items that can picked and chosen from to fit the particular taste of an operation. There is an order in which they should be applied. For the most part, we haven't done well in our choices because what has transpired thus far hasn't resulted in fully transforming manufacturing and collectively achieving world-class advances in waste reduction, production flexibility, and customer satisfaction.

The recent shift to a well-defined payback on investment goes against the grain of the just-do-it philosophy utilized in the early stages of Lean. But in all fairness, three words should be added to the cliché, making it: *Just-do it — but with knowledge.* Others can strive as hard as they like to make the case that Lean Manufacturing has been everything that it's touted to be, but the reality is we've been focusing on the wrong things. As a result, management hasn't seen the kind of results that would unquestionable justify a much more aggressive pursuit of the process.

Correcting the problem centers on a plant's key production equipment and making certain that it is properly "geared" to support the effort. Accordingly, the ideal situation would be to apply what will be pointed out to operations that have not, as yet, started a

Lean initiative. This isn't to say that a correction in course should not be made by those who already have a Lean initiative underway — such efforts have the potential to greatly enhance the implementation of Lean Manufacturing in any factory, regardless of size or the type of products produced.

INTRODUCTION

To a large degree, Lean Manufacturing in the United States has been slow off the starter's block and has gradually shifted to a cost-based process, hidden under the guise of continuous improvement. This leaves an important question to ponder: *Does America choose to use Lean to its fullest and significantly enhance our competitive position in the world, or do we continue to focus on the low hanging fruit, in lieu of rejecting the primary mission of the process?*

Although there are many types of Lean methodology on the market, each with its own particular slant about how to go about the process, they all incorporate the tools and techniques founded in the Toyota Production System. Thus, the issue isn't which method is the best. The issue is clearly understanding how to go about *sound implementation strategy and where, when, and how to apply the tools to a company's best advantage.*

If we look closely at what's transpiring with Lean across the United States, what we typically find are islands of improvement that dot an ocean of waste and inefficiency. Such results simply will not take us far enough (and fast enough) to stem the growing tide that is currently serving to erode the U.S. manufacturing base.

The ideal situation, of course, is when leaders recognize that the need for Lean is so great they're willing to set aside other initiatives until the task is fully and successfully accomplished at a factory level. Because this type of commitment is seldom, if ever, the case, we have to look at how to go about implementing Lean in the least disruptive and most effective manner. In order to make this happen, we have to get past the growing mindset that strives to implement the process through small incremental improvements.

Instead, we must begin to focus on a larger objective — which is to make a full and complete transition across the entire production arena and the general supply chain.

Within the past five years, a growing trend has been placed on incorporating Six Sigma, a methodology developed by Motorola in the mid 1980s. This methodology was initially aimed at improving product quality by identifying and correcting variations in processing. But the truth of the matter is although Six Sigma is a good tool which can often provide solid results, it simply isn't what Lean Manufacturing needs as a *driving strategy*. In fact, in many ways, it has served to place focus on absolutely the wrong strategy.

In a 2008 article from the Six Sigma Academy, outlining the benefits of the process, the following was noted: *"Black Belts save companies approximately $230,000 per project and can complete four to six projects per year. The Six Sigma process (define, measure, analyze, design and verify) is an improvement system used to develop new processes or products at Six Sigma quality levels. It can also be employed if a current process requires more than just incremental improvement. Both processes are executed by Six Sigma Green Belts and Six Sigma Black Belts and are overseen by Six Sigma Master Black Belts."*

I would have personally felt better if the improvements had been outlined in terms of processes fully transformed to Lean. But as a counterargument to the insinuation that Lean's objective is cost savings, a couple of very important points should be made that are pertinent to the content of this work:

1. No company can persistently save its way to profitability. When it's all said and done, it isn't a matter of how much money a change can save a company that counts. It's how much money it's going to *make* the company over the long haul. Once a company clearly understands that the objective is to work energetically at spreading the force of a well-established and proven process across the entire manufacturing arena, in the fastest manner possible, they can start to make some remarkable strides in improving their overall effectiveness and profitability.

2. The driver for a good Lean initiative has to be the firm belief that in order to compete with the likes of Toyota and others, the existing system of production has to be *fully destroyed* and replaced with a far less waste-infested means of doing business. If done properly, this change will almost always require making an initial investment, as opposed to achieving a cost savings.

One of the best ways to slow the thrust of a Lean Manufacturing initiative is to require some form of evidence as to Return on Investment before change of any kind is allowed. Firms that are good at Lean understand that the people implementing the process should follow a set of guiding principles. In doing so, as long as the change clearly meets the criteria for the principles noted, no other form of justification is warranted. In other words, "just do it," but with a knowledge that allows the focus to be properly placed.

Irrespective of philosophy, placing everything on the altar of Kaizen has served to hinder progress more than inspire it. In order for us to gain parity with the likes of Toyota and others, executive management has to accept no less of an excuse for not swiftly and fully implementing Lean Manufacturing than it would for not fully satisfying customer requirements or providing products of good quality. In turn, plant management has to see Lean as an absolute requirement and reporting functions have to perceive it as a top priority. Doing this boils down to approaching Lean with the same level of zeal and commitment Toyota did in its infancy. However, this type of commitment isn't happening on a broad scale in the

United States. Anyone who believe otherwise is vastly misleading themselves.

Best put, implementing Lean in the most effective manner boils down to doing the *right* things, at the *right* time, with the *right* people involved. Therefore, with an expressed emphasis on both the speed and smoothness of implementation in mind, this work is designed to focus on what has served to deflect energy. It will point out the lack of appropriate expertise commonly applied to some of the more important tools of Lean and how implementation strategy should be adjusted accordingly. Key to doing this will place focus on three major components of implementation:

- How driven a factory's shop floor leadership is about the process
- How qualified the dedicated resources are in applying the tools
- How geared a plant's equipment is toward supporting the process

HOW TO MAKE THE MOST OF THIS BOOK

In the leading section of his book *How To Measure Managerial Performance,* Richard S. Sloma pointed out, *"I will have failed to achieve my most important objective if all you have done is merely to have read this book. We will both achieve our objectives only if you USE this book."*

I feel much the same way. Respectfully borrowing from the technique Mr. Sloma employed, I would like to point out how to make the most of this work, considering there will be both readers who are new to Lean and those who are seasoned in the tools and techniques. The book contains something of importance for both because the principal topic deals with implementation strategy.

The first section (Chapters One and Two) provides a comprehensive overview of the current path Lean is taking, along with the myths, misconceptions, and flaws that have hindered progress. These are the areas where attention needs to be focused in order for America to be much more proficient at Lean. Four levels of Lean Manufacturing are outlined which can be used to measure progress.

The second section (Chapters Three through Five) deals with what the war on waste is all about, the organizational side of the equation, and various levels of measurable accomplishment. Chapter Three can be viewed as a buffet of items from which the reader can pick and choose from in reference to a specific topic. But it should be noted that although this book is principally directed at those who have not as yet entered into Lean, it can be an excellent refresher for those who have a Lean initiative underway. Chapter Four moves on to address the important topic of properly organizing and aligning personnel for the effort and in understanding how various leadership styles both add and distract from implementing Lean in an effective manner. Chapter Five, in turn, centers on the details of how to achieve a Level One Lean Status and measure overall results.

The third section (Chapter Six) deals with the more advanced aspects of Lean Manufacturing and covers such topics as the importance of an 18-Month Rolling Implementation Plan and making a Core Process Analysis; the reasons and benefits for those who are already into Lean to strongly consider a change in course; how to gain support and cooperation from major supporting functions such as Accounting, Sales and Marketing, along with other important topics. At a minimum, this section should be scanned initially and picked up and read again, once a plant has reached a Level I status.

At the end of each chapter (including the Introduction), a number of "Key Reflections" associated with the text are noted, for quick reference purposes.

KEY REFLECTIONS

- Resting everything on the altar of Kaizen (making small, continuous improvement) has hampered more than inspired progress.
- The issue of equipment that isn't geared to support Lean is seldom, if ever, seriously addressed on the front end of most Lean initiatives.
- The United States has proven repeatedly, if and when it puts its heart and soul into an effort, it has the ability to deliver. But for various reasons we haven't as yet established a universally accepted means of implementing Lean and effectively measuring progress. This, in turn, has left many organizations questioning where they actually stand and what they've truly accomplished.
- The best intentions on the part of manufacturing managers and first line supervisors can't be readily applied if we're asking them to fight a battle with one arm tied behind their back.
- The answer rests in a process of implementation that guides an operation to doing the *right* things, at the *right* time, with the *right* people involved.

THE BASIC FLAWS AND
MISCONCEPTIONS ABOUT LEAN

Batch manufacturing has been both the principle factor for success and the nemesis for a decline in the dominance of American manufacturing. Growing out of the techniques employed by Henry Ford when he strove to make an affordable automobile for the masses, batch manufacturing became the principle reason the United States was viewed worldwide as a model for industry. But as time prodded on, it was to become the chief reason the United States lost the leadership role it held in manufacturing for well over six decades.

During World War II, batch production was refined to as a science; it helped establish a "more is better" mindset. But that alone probably would not have made this system of production the waste generator it is today. Starting in the mid-1970s, the average consumer was exposed to an ever greater offering of styles, functions, and designs. Today a high level of product diversification is fully expected. The downside was this approach significantly increased the need for more on-hand inventory, added more equipment that had to be maintained, and created substantial variations in processing — all of which served to grow the wastes noted to monster proportions.

But it wasn't until the late-1980s, when Toyota and other Japanese manufacturers began to invade the scene that U.S. manufacturers came to discover they were facing an opponent who went about the task of manufacturing in a totally different manner. Out of arrogance on one hand and a serious miscalculation on the other, the need for change was principally ignored, until it could no longer be denied that the United States was starting to lose its manufacturing base.

What made matters worse was even though most firms came to confirm the need for Lean Manufacturing in the mid-1990s, we haven't done an adequate job of fully and effectively implementing the process across a broad spectrum of industry. Thus, the purpose of this book is aimed at how to go about *completely destroying batch manufacturing* and replacing it with a world class system of production, in a reasonably fast and effective manner.

A question that comes to mind is: Are we looking at setting aside everything we're learned about Lean and starting over again? The answer to that is contained within the content of this work. I can assure the reader, however, that developing a world class system of production doesn't require setting aside what has been learned in the past. Nor does it require starting over again. What it boils down to is revising our implementation strategy, especially where the sole focus has for the most part been a shotgun approach to continuous improvement. But in making changes, nothing has to be undone. The best way to view the effort required is greatly enhancing the work that's been accomplished thus far.

IMPLEMENTATION STRATEGIES

Meaningful ideas aren't created in a flash of brilliance. They're born as a result of first-hand experience, a sense of dedication to the process, and a willingness to step forward, even when it goes against the grain of traditional thinking. As we move forward, let's consider some common perceptions that have been formed with respect to implementation strategy.

Perception #1

The mission of Lean Manufacturing should be to make steady, incremental improvements that result in an immediate savings to the company.

Counter Argument

Making Lean a success requires a focus on changing a factory's entire approach to manufacturing. If done correctly, this change requires spending money on the front end; this investment most often will not result in an immediate payback.

Perception #2

Implementing Lean does not require additional staffing, outside of perhaps hiring an experienced Lean Coordinator to oversee the process, perform training, and track overall results.

Counter Argument

While efforts are being directed at implementing Lean, the factory is still working to meet customer requirements under the rules and operating guidelines of the existing system of production. As a result, the ability to shift roles and responsibilities within the current ranks

is extremely limited, and usually provides less-than-adequate support for Lean.

Perception #3

Factory participants with no real experience and background in engineering can be trained to successfully conduct work measurement and to effectively apply the sciences of setup reduction and mistake proofing.

Counter Argument

Performing dependable work measurement — critical to operating decisions pertaining to manpower, taking on added business, and the like — requires expertise in performance rating and methods evaluation. This expertise is inherent to the science of Industrial Engineering. In addition, conducting meaningful setup reduction and working to make production processes mistake proof, which is critical to the fundamentals of Lean Manufacturing, cannot be effectively accomplished without the skill and expertise of Manufacturing Engineering.

Other Perceptions

There are other misconceptions about implementing Lean that will be addressed as we move along. One of the principle factors, however, has been the unerring belief that Lean Manufacturing is a never-ending process. Therefore, if it took Toyota four decades, it's perceived that the United States will need a similar portion of time to gain the same competitive level of expertise. As a result, the overall expectations of management and stockholders alike have been minimized. The pressure to find a means of implementing Lean in a quick and effective fashion has fallen by the way-side.

There is a very serious flaw in this type of thinking. Fully implementing Lean Manufacturing is *in no way* a never-ending process. What is never ending about it is the continu-ous improvement aspect, which can best be done when a plant has made a full and absolute change to its existing sys-tem of production.

A very frustrated hourly worker pulled me aside once to complain, "It shouldn't take years to do this. If they can move a factory to Mexico and have it up and running in nine months, we should be able to make the same kind of change we need here, in order to help keep our jobs." When I passed the comment on to his direct supervisor, his reply was, "Freddie's a good guy, but he doesn't understand we've got other things to do as well." I proceeded to ask why Lean was-n't being more aggressively applied in the factory. He summed his answer in one word, "Equipment." When I point-ed out I'd been informed the equipment involved had setup reduction and mistake proofing applied, he responded with a sly smile, before adding "For all the good that's done! They really shouldn't have wasted their time."

I've since come to believe what he really meant to say was that it was impossible for him to be enthused about Lean when everything he had to work with was geared to accom-modate a totally different style of production. We have to understand that the best intentions of our production man-

agers and supervisors cannot be readily applied if we're asking them to fight a battle with one arm tied behind their back.

Lean Objectives

For every case of striving to introduce Lean and
spread it across a factory, there are two objectives that
always take priority. The first has to do with meeting established demand, by achieving a master schedule that is intended to reflect actual customer orders. The second objective is
to achieve assigned budgets and forecasts, without incurring
unfavorable variances.

Lean is touted as a process that shouldn't pose a significant interference in meeting customer demand, along with
projected budgets and forecasts. But it frequently can and
does. When implementing Lean seriously interferes with past
practices, it can be perceived as more of a nuisance than a
benefit. And when such a perception sets in, the process is
doomed to become a secondary mission, with little heart and
soul behind the effort.

What is simply amazing is the number of manufacturing leaders who believe that achieving a complete transition
in production can be done on a part-time basis and with limited and less-than-qualified resources.

Equipment Engineering

Toyota created a substantially different approach to the
mass production techniques used by American industry at the
time. Their approach led to the *Toyota Production System,*
which has since become the foundation for almost every initiative undertaken in the field of Lean Manufacturing. But

there is an important piece of the puzzle that's been lost: a plant's Key Production Equipment.

Any difficulty in moving a Lean initiative forward will almost always revert back to equipment that isn't geared to respond to the conduct required. I use the word *geared* in reference to the compliance of equipment to the highest levels of both reliable and repeatable performance. Equipment that can achieve these levels will lead to improved flexibility, substantially lower operating costs, and an often eye-opening reduction in factory lead-time.

The issue of equipment engineering seldom, if ever, is addressed seriously on the front end of most Lean initiatives. As a result, the process has been hampered. Consider if you will, the following questions:

1. Why do firms enter into Lean with nothing but good intentions, only to see it seriously falter before it becomes a way of life?
2. Why hasn't America's Lean transition become an answer to outsourcing hundreds of thousands of manufacturing jobs to foreign soil?
3. Why is it so hard for U.S. industry not only to buy into the concept of Lean Manufacturing, but also to use the nation's considerable ingenuity in making it a reality?

The United States has proven repeatedly we have the ability to deliver, if and when we put our heart into an effort. But for various reasons we haven't as yet established a universally accepted means of implementing Lean and effectively measuring progress. In turn, this failure has left many organizations questioning where they stand and what they've truly accomplished.

Take the U.S. space program as an example. Everyone would agree that it too is a never-ending process. But a very narrow focus was placed initially on putting a man on the moon. That goal, in essence, became the United States' space program and the nation rallied around it with an uncommon fervor. However, meeting this objective was done with the expressed knowledge that while the program would not effectively *end* with that achievement, it could not truly *begin* without it.

The same holds true for Lean Manufacturing. We must have a recognized level of accomplishment, at some well-defined point in the process, in order for manufacturers to buy wholeheartedly into the concept and work to achieve a foundation from which the next solid commitment can be launched.

But first one has to acknowledge that if a factory can never look someone in the eye and say they've arrived, so to speak, there is little chance they will have the fortitude to push forward aggressively toward the next level of achievement. As a result, the overall mission requires a spot where an operation can pause, if only momentarily, to celebrate a clear and noteworthy accomplishment — not an end in itself, but rather an achievement that becomes the next launching pad for the future.

OBSTACLES TO PROGRESS

Before examining what can done to improve the thrust of implementation and more quickly gain the benefits across a broad scale of U.S. manufacturing, it is important to summarize the flaws that have served to hamper progress:

1. The lack of an appropriate focus on a plant's key production equipment in setting the stage for an aggressive application of Lean across the entire operation.
2. A general failure in the utilization of the Industrial and Manufacturing Engineering functions in the process.
3. A growing trend away from a "just do it" mentality to establishing a proof-based comfort level before change of any kind is allowed.

After retiring, I was called on by various firms to assist in setting the foundation for a *Waste Free Manufacturing* environment. In every case, some very pronounced accomplishments were made. Work-in-process inventory levels were reduced as much as 90%. Productivity, in terms of the actual number of people required to perform the work, improved as much as 30%. Required floor space was reduced up to 50% and quality measurements, in the form of scrap, rework, and obsolescence, were lowered 50% and more.

Under any form of evaluation these would have to be classified as phenomenal accomplishments, especially considering the change was made over a very short period of time. On the other hand, there was more than one occasion where completing implementation on a plant-wide basis fell short of the goal. Although it would have been easy to say there simply wasn't strong enough management support, that wasn't the case. Management was more than willing to see

Lean become a success and to fully support it — up to a point. That point, of course, was when Lean began to seriously distract from achieving other factory obligations, such as dealing with expenses and meeting customer demand.

There are those, including myself, who would like to see plant management much more driven as to the need for Lean and more willing to step forward in defense of the process. Still, we have to face reality. In the vast majority of cases, this isn't something that can be depended on to keep a Lean initiative at the forefront of priorities. Consider the case of Avery Manufacturing (Case 1-1):

CASE 1-1 AVERY MANUFACTURING

Avery Manufacturing, which has been in business for well over two decades, produces plastic extruded components for the automotive industry. For much of its existence, it enjoyed steady growth and improved market share. But as competitive pressures grew, it slowly began to lose business to overseas competition. As a result, profits and share of market began to spiral. After much deliberation, management decided there was a need to pursue a Lean Manufacturing initiative.

After communicating to employees, Avery hired the services of a well-respected consulting firm. As a first step, an area of the factory was selected as a pilot project. A special event was conducted involving a number of key factory personnel, including the plant manager and various members of his staff.

The event went extremely well. Participants received training in the basic tools and techniques. As is usually the case, the chosen pilot area was totally revised. Floor space was reduced, required work-in-process inventory levels were lowered, direct labor was redistributed, and manpower adjustments were made. Unneeded items consisting of inventory, old and infrequently used equipment, and such were removed from the area and stored in a special zone until a decision could be made as to disposition. In addition, work stations were redesigned with input from the operators; numerous visual controls were installed.

Afterwards, enthusiasm ran high. Work began on spreading the change plant wide. Twenty-four months later, however, one could find little evidence of a successful turnaround. Factory inventory levels remained as high as ever and slippage was evident in the selected pilot area, especially regarding work-place organization. Although a substantial number of smaller in-house events were conducted after the pilot, focus had been placed on making small improvements within the confines of larger production departments, which tended to be suffocated by the batch environment going on around them. As added competitive pressures grew, more and more effort was shifted from implementing Lean to addressing and resolving immediate production issues (firefighting). The strong enthusiasm on the front end slowly began to ebb and largely turned to skepticism on the part of employees. They began to view Lean as just another program, among the many that had started and died over the years.

This case is, for the most part, a fictional account. But it points to what's transpiring in much of U.S. industry. Initial efforts are generally impressive and filled with unique accomplishments and high enthusiasm. Following this, however, things frequently begin to slow, principally as a result of not fully understanding what to attack first, second, and so on (see Figure 1.1).

Figure 1.1
How to Go about the Job

✔ **Establish clear levels of accomplishment:**
   Level I through Level IV*

✔ **Determine the tools needed:** Poke-Yoke,
   TPM, SMED, etc.

✔ **Train and communicate**

✔ **Enlist the workforce**   * Details spelled out in
                                     Chapter Two, Figure 2.2

To emphasize what I'm driving at, I once worked with a well-known firm where, six months after a highly successful event, I returned for a follow-up review. I was astonished to see that outside of some rather insignificant changes on the factory floor, little progress had been made. In addition, the pilot area, which was designed to be a showcase for how the process should both look and feel, had shifted back to a push system of production, after initially being targeted as the first pull area of the factory.

Upon further investigation, it became apparent that the objectives established for the change effort had in no way been met. In fact, no machine in the factory had a setup time less than twenty minutes in duration and some machines took hours to change over. On two projects where team members had placed some effort, the post-pilot goals for setup reduction were far from achieved and no work whatsoever had been applied to error-proofing equipment. Even more disappointing, I learned in a follow-up meeting with plant management that they were pleased with the work accomplished. They noted that although the goals hadn't been fully achieved (a vast understatement), the team had improved setup on two pieces of equipment.

Much of their response was an effort to justify where progress stood, in order not be seen as lacking in their commitment. But as politely as I could under the circumstances, I cautioned them that the degree to which they expressed satisfied or disappointment said a lot about where they ultimately intended to take the process.

The silence was almost deafening as I told them that I didn't think Lean was really all that important to them. The plant manager, in particular, was visibly upset and asked me to provide the reasons I felt that way. In response, I proceeded to give each of them a copy of the participant feedback form I have team members complete on a follow-up visit. Among the findings:

1. No meetings had been conducted by management to check on how things were going or to redirect the activities of the team as needed, in achieving their stated objectives.

2. Collectively less than eighteen hours over a six-month period had been made available for team members to

work on stated objectives.

3. Although a majority of the team believed management thought Lean was important, all of them noted that "other things" came first, including:
   - Meeting production schedules
   - Meeting forecasted operating expenses
   - Providing support to higher priority or more important plant and corporative objectives

I noted that anything more than single minute changeover fell short of World-Class. It wasn't insignificant to the decision making process for issues such as adding business, increasing line rates, etc. I further reminded them that other pressing matters and higher priority objectives will always be there, in one form or another. In order to move a Lean initiative forward at a reasonable level of speed, there has to be a commitment to dedicate some number of resources to the process on a full-time basis, or at minimum some pre-determined period of time.

I should note that management was in no way disinterested or thought that Lean was less important than other things. They were simply typical manufacturing managers, working under typical conditions, which strongly influenced an operating mentality that said:

- "Things are always going to get in the way, so never overstate an objective. If anything, strive for a goal that's something less than possible and offer a pat on the back for any improvements made."
- "The most important thing is to keep banging out parts

and components, even if it takes an abundance of downtime, scrap, and rework — and if and when inventory becomes an issue, we'll take our limps and move on."

The problem many manufacturing managers have is that they simply refuse to get out of the way of progress. They do not believe machines can run without breaking down and without producing scrap and rework. They do not believe setup can be reduced to near zero and that errors inherent to specific pieces of equipment and processing can be entirely eliminated. What they do believe, however, is there's no magic that would serve to make manufacturing anything other than a day-to-day chaotic exercise. Otherwise, they'd be pushing the hardest for the change and, in most cases, would be staying after hours and weekends to make it happen.

Admittedly, implementing Lean puts a strain on expenses, drains needed resources, creates unneeded downtime, and for the most part has no immediate impact on the big picture. But place the initial thrust on effectively improving a plant's key production equipment, which for years has served as the one thing that poses the greatest stumbling block to achieving Lean's stated objective, and attitudes will shift dramatically.

The Japanese and more specifically a number of ex-Toyota managers were the first to bring the general philosophy of the Toyota Production System to U.S. shores. The thing they never seemed to clarify, however, was precisely what should come first, second, and so forth, in order to move the process across the entire factory. There could have been many reasons for this, including the possibility they simply didn't look at it in those terms. The skeptic, of course, would say it wasn't in their best interest to show the United States how to gain parity. I lean toward the theory that they didn't view the process in terms of speed of implementation, but rather in making certain that participants understood how the various tools and techniques were intended to work.

Anyone who knows anything about Lean Manufacturing has a special admiration for Toyota and what it accomplished. They have served as the basic role model for Lean initiatives in the United States. But suppose Toyota was placed in the position of having to do it again. Would they take the same basic steps we're using to implement the process?

I posed that question to a number of people who were implementing Lean in various organizations; they generally had to think about it a bit because it was something they had never considered. The majority came to the conclusion that Toyota would follow the same path we are currently using. Those who didn't respond in like fashion admitted they really didn't know for certain. No one was convinced Toyota would go about it in an entirely different manner.

I believe if Toyota had to do it again, they would first gear their equipment to support Lean, through a highly professional application of SMED (Single Minute Exchange of Dies) and Poke-Yoke (a Japanese term related to mistake proofing equipment). In fact, Toyota applied much more attention to their equipment than has come to be recognized — not because they were striving to hide something from us, but because we did not pay close enough attention to what the recognized father of the Toyota Production System, Taiichi Ohno, was trying to tell us.

Implementing Lean

With respect to this, my entire thinking on the path of implementation was changed in the spring of 2006, as a result of being asked to consult with Brunswick in the development of a Corporate Lean Manufacturing Guideline. The workshop was conducted in one of the company's feeder operations in Mexico, which served as an internal supplier of wiring harnesses.

I was privileged to see a factory that had made outstanding progress. During the session, I was asked to rate the operation on a scale of one to ten. My immediate response was that the plant was a solid nine. But it quickly became apparent the group of executives in attendance was seriously disappointed I hadn't answered a ten.

I explained that my assessment was based on what I outlined in my first work *Fast Track To Waste-Free Manufacturing*, pertaining to three established levels of achievement. Although the plant was very close to reaching what was outlined as a Level III, there were still some things left to be done, such as the insertion of Owner-Operators. I did clarify that the plant was unquestionably one of the best I'd witnessed in terms of a plant-wide application of the tools and techniques.

One of the more compelling features of the factory was the speed in which continuing change could be made. For example, during a morning tour of the factory, I mentioned to the plant manager that he might want to place center locators on press beds for quicker insertion of dies and fixtures. During the afternoon break, he showed me a machine that had been fitted with the locator and pointed out that the dies had been notched accordingly; all of these adjustments had to be done from scratch in the plant's machine shop.

I congratulated the manager and politely stressed that putting such an effort into making the change was admirable, but showing me that he was going to use the idea wasn't necessary. I was already highly impressed with his factory. He look at me strangely a moment before replying, "Oh no, we would never make a change for show. Improvement to our equipment always comes first, before anything else."

It was then that I realized precisely how the factory had made so much progress, in such a short period of time. Their principle thrust, inadvertently or otherwise, had been to apply Lean Engineering to their equipment, In other words, they insured from the start that their equipment fully supported the process, in terms of an extensive and highly professional application of setup reduction and mistake proofing. They went further by keeping this in the forefront of their continuous improvement efforts.

As a result, it became clear to me that any operation, no matter how entrenched it might be with a batch mentality, could more effectively insert Lean if it placed an initial focus on getting its *key production equipment* aligned to support the effort. Something I can say for certain is if I had the chance to do it again, at any of the factories I was responsi-

ble for turning around, I would definitely take this approach —
not because what was done was a failure by any means, but
because I'm solidly convinced this approach would be a better way of handling the task.

I would challenge anyone to imagine how much easier
it would be to make fast and lasting change on the shop floor,
if a factory's key production equipment had a highly professional level of SMED, Poke-Yoke, Standard Work, and TPM thoroughly applied *before* starting to make flow changes on the
shop floor. This doesn't mean other important elements of the
process, such as Workplace Organization, 5-S, or 6C, could
not be incorporated at the same time in many areas of the factory. But it does mean that work aimed at changing the flow,
establishing point-of-use manufacturing, and setting up supply
and replenishment links would be deferred until a plant's key
equipment has been fully engineered (see Figure 1.2).

I have found that what is commonly viewed as the
more obvious objectives of Lean are often seriously compro-
mised because of production processes incapable of adequately supporting the change. As a result, the process tends

PRIMARY FOCUS	SECONDARY FOCUS	DEFERRED FOCUS
• SMED • Poke Yoke • TPM Where initial work and training	• Workplace Organization • Standard Work • Visual Controls Training and work that can run concurrently, as time and resources allow	• PT. OF USE • KANBAN • PULL PRODUCTION Training and work that is deferred until the Primary and Secondary focus are fully completed

Figure 1.2 Training Stages and Focus

to lose the muscle it needs to move forward at an aggressive pace.

Toyota was successful because they were never satisfied with any single achievement or series of accomplishments. Instead, they were focused on a mission of total waste elimination. Completing this mission meant preparing the stage accordingly. To prepare, they had to get production processing as close as possible to zero setup, and also eliminate common processing errors. When a plant's equipment approaches this level of applied engineering, there is little excuse for the process not to move forward in an aggressive manner.

The overall content of this work provides the framework for what I sincerely view as a more appropriate means of implementing Lean, regardless of the size, type of operation involved, or products produced. Examining the key factors that have led us to where we are today is also important. In that way, we can understand precisely what we need to do differently.

THE TOYOTA FACTOR

There can be no argument that Toyota left a lasting and vastly important impression on Lean efforts in the United States. Read any book on Lean and you will find reference after reference to Toyota, along with examples of the improvements they have been able to bring to manufacturing. But nowhere to my knowledge can you find a specific step-by-step outline, as to the path that should be taken in implementing the process. Without this path, we have had to make our own way; as with any process of trial and error, there have been both successes and failures. But we haven't, as yet, adopted a proven and universally accepted method of going about the task.

To strengthen the point, let's re-examine the typical technique used to launch Lean Manufacturing in the United States. In this basic approach, management appoints someone as its Lean guru. Management often contracts with a qualified consulting firm in order to get the ball rolling. A given production area is selected. Then over a period of time, participants are provided training in the various tools and techniques; they are asked to make what often amounts to eye popping change on the shop floor.

Not so surprisingly, this normally turns out to be a highly successful venture. Everything from work station layout to simplified replenishment techniques are addressed and improved. Furthermore, new shop floor inventory levels are established, operator work is redistributed, and the pilot area is generally assigned new levels of throughput. As a result, some very substantial improvements are generally gained. In addition, the area is cleaned and shaped, and ends up performing in a much more disciplined manner than ever before.

But after the usual celebration, what we typically see is that the pilot area slowly begins to drift back to doing business as usual. If we were to look deeply into what's actually transpiring, we would find that much of the noted stall boils down to production equipment that isn't geared to support the process. As a result, newly established inventory and throughput levels become theoretical at best The pilot area begins to steer in a direction opposite to the fundamental principles that were initially applied.

As mentioned earlier, many of the top consultants at the time were ex-Toyota managers. They were the first to bring the United States an abbreviated version of the Toyota Production System. As a plant manager, I was as inspired as anyone. Over an eighteen-month period, I went on to literally force the change in a factory that became a showcase. Later, a number of Wall Street analysts visited the plant and reaped considerable praise on the process. But the critical point to be made is that the abbreviated application of the Toyota Production System became the basic model for almost every Lean Initiative undertaken in the United States.

The Japanese were focused on quickly exposing participants to the basic tools and techniques. On one noted occasion, the approach went so far as placing an entire final assembly line in the company's packing lot — then over the weekend, rearranging work stations followed by fully reassembling the line on the shop floor. You can imagine the chaos that followed, when operators returned the following Monday, to find everything about their world had totally changed. All this was aimed at getting the workforce's attention regarding both the magnitude of change required and the potential benefits to be gained.

What the Japanese didn't dwell on, however, was precisely how to conduct a full insertion of the process across the entire factory. Without trying to put words in their mouths, they perhaps felt, "We can show you how to fly the airplane, but you have to be your own pilot." We should keep in mind that the techniques were essentially bundle implemented over an extremely short period of time. This was necessary in order to make the point regarding precisely how the tools were intended to work. On the other hand, the system of production was developed and deployed over four decades at Toyota. We should not take for granted that the flavor presented by the consultants was precisely how to go about successfully installing the process, factory wide.

But again, the approach taken by the Japanese became the model most U.S. companies used in both starting and moving the process forward. In most cases, continuous improvement became the by-word and the ultimate measuring stick for success. As long as some improvement was being made, no matter how small or convoluted the effort might be, working in this manner was generally perceived as how things were intended to proceed.

Approaching Lean in this fashion is one way of getting there. However, it's much more time consuming and open to some very serious stumbling blocks as the process trudges along. There are four questions every good manufacturing manager who has been into Lean for a time should ask:

1. Are you truly satisfied with the progress of your company's Lean initiative?
2. Are you seeing solid profit improvement or something short of what you perceived the process would yield?
3. Have manufacturing lead times and overall inventory levels been significantly and permanently reduced?
4. Are your customers going out of their way to let you know how much more satisfied they are with delivery and product quality?

Any firm entering into Lean should strive to be as effective with the process as Toyota, Komatsu, Yanmar, Hatachi, and others. Anything less would be short sighted. Unfortunately, most firms never reach that level of commitment. They tend to approach Lean in a rather disorganized manner; in lieu of no clearly established end objectives, the process tends to wander. Although continuous improvement is is indeed vital to the process, much of our industry is striving to position itself where Toyota currently stands after forty years of effort. The only difference is Toyota is working to improve processes that have been highly engineered; for the most past, we aren't.

The book *Kaisha: The Japanese Corporation* by James C. Abegglen and George Stalk Jr. notes the following:

A Japanese factory required about thirty days to make its product before the development of JIT. Once setup times throughout the factory were reduced, the production period fell to twelve days. After the layout of the factory was changed to reduce material flow and inventory holding points, the production period was reduced to six days. Eventually, as all inventory holding points were eliminated, the period of production was reduced to two days. This is the equivalent of reducing the production lead time of the European construction equipment company from twenty-four weeks to a week and a half.

It is important to note that in this general overview of Japanese manufacturing, the very first thing done, and the one that had the single greatest impact on lead time (an 18-day improvement) was placing an initial focus on reducing setup. Nothing was mentioned about continuous improvement being vital to what was accomplished, leaving the impression that the mission was to make it happen as fast and effectively as possible.

The point about adequate equipment engineering can be magnified by looking back at what Taiichi Ohno, the founder of the Toyota Production System, had to say in 1988, when asked where Toyota was headed because they "must have reduced all work-in-process inventory, lowering the water level to expose the rocks, (and) enabling them to chip away at the problems."

Ohno's response was, "All we are doing is looking at is the time line, from the moment the customer gives us an order to the point when we collect the cash. And we are reducing that time line by removing the non-value added wastes."

Ohno indicated that Toyota had reached the point where it could successfully apply continuous improvement. What he didn't say, but which I wholehearted believe, is that this should only become a driving objective *after* a great deal of fundamental work has been accomplished on the shop floor — more specifically, in the arena of engineering equipment to support the process.

The level to which Toyota currently operates came over the course of 26 years of development and another 18 years of improving the process. We can avoid a great deal of trial and error by focusing on the techniques developed by the Toyota Production System. In doing so, however, we have to recognize that some fundamental steps must be taken, and be wise enough to see that they are fully accomplished. This calls for understanding what the operating rules should be, along with the steps required to implement the process, in a timely and orderly fashion.

THE COST JUSTIFICATION FACTOR

As I have said, Six Sigma is an excellent tool and I highly recommend its use. The problem is in the way it is applied in most cases and, just as important, the role it has played in creating a false perception about how to implement Lean.

Something I offer in my consulting business is a free factory Lean assessment. I offer the assessment for two reasons. The first is to help factories verify they're on the right track and to establish where further opportunities for improvement exist. The other reason is it allows me to continue to expand my knowledge and experience. The assessment provides me a means of keeping up to date with industry developments.

During a recent assessment, conducted in the fall of 2008, I was informed in the initial meeting with plant management that the operation had four Six Sigma Green Belts, two of whom were close to achieving Black Belt status. The ensuing conversation went something like this:

"So what kind of things do you have them working on?" I asked.

My question was met with a look of puzzlement. "Oh, just typical Six Sigma projects," the answer finally came.

"What I'm driving," I strived to clarify, "is where the

basic implementation strategy is being directed?"

"Their efforts are being applied across the board and we're seeing some excellent improvements," I was informed, before he went about showing me a list of projects that had been achieved over the previous twelve months and the pay-back on investment.

In quickly looking over the list, I saw a couple of things that fit well with implementing Lean. However, I saw far more items which were basic cost savings or cost avoidance projects, such as changing material in order to lower a purchase price, and eliminating redundancy in the order entry process. I decided to put the subject aside for the time and learn more about what was actually transpiring on the shop floor. What I discovered served to reinforce what I've seen in many U.S. factories.

The plant had done some reasonably good work in the basics, especially in the area of workplace organization and establishing various areas where the principles of U-Cell flow were being utilized. But no work whatsoever had been performed on setup reduction. I found that interesting because 20% or more of an operator's time, on average, was spent setting up the equipment.

I'm pleased to say the result of the assessment led the plant to make a correction in course. The power of Six Sigma was defused, to a large extent by projects that had little to do with aggressively implementing Lean Manufacturing. This isn't to say what they were working was of no significance or importance. But with respect to the training and leading role assigned to this particular resource, their efforts were simply misdirected.

Just as important as the issue of misdirection is the perception this type of approach has created on an almost overwhelming scale in U.S. industry. Lean, to a large extent, has steadily transformed to a project-based process, with clear cost justification required before change of any kind is made. Nothing could be further from the mindset and conduct needed to implement Lean Manufacturing fully and effectively.

What needs to be done differently is to provide those performing Six Sigma with some reasonable flexibility in moving forward. We should be teaching them the guiding principles and allow the principles to guide their actions, without undue restrictions. Coupling this teaching with good implementation strategy, aimed at getting a plant's key production equipment in tune with the effort, would unquestionably result in some truly amazing accomplishments for U.S. industry.

THE ENGINEERING FACTOR

Another important issue that has played a role in the lack of overall progress has been the placement of less than qualified resources in the pursuit of three key elements of Lean: Setup and Changeover (SMED), Mistake Proofing (Poke-Yoke), and Work Measurement (Standard Work).

The matter of applied engineering has been seriously downplayed in the United States, principally due to the perception that came out of the early Japanese influence on the process. I have repeatedly seen the process falter after the initial stage of implementation, most of which has occurred because proper expertise was not applied to SMED, Standard Work, and Poke-Yoke.

Given reasonable training, many of the techniques of Lean can be carried out by participants who have no engineering expertise; such as 5 S, visual controls, and the like. On the other hand, SMED, Standard Work, and Poke-Yoke are extremely technical tasks that are best suited for the talent of a seasoned engineer. In fact, I would go as far to say, they cannot be adequately and thoroughly undertaken otherwise.

Ohno spoke to the appropriate use of engineering talent. Although his words have served as the foundation for much of what Lean Manufacturing has come to be in the United States, the engineering aspect of the process has been sorely under-emphasized. In fact, I believe some of his words have been taken completely out of context. To make the point, we should examine some facts, misconceptions, and conclusions.

Fact #1

In the vast majority of undertakings, Standard Work is performed by hourly and salaried participants who have no real experience and minimal training in the science of Work Measurement.

Associated Misconception

The theory behind this approach is if the person being observed is an experienced operator, there can be reasonable assurance the operator will perform with good effort and to the prescribed method.

The Problem

There are several problems with this, two of the more important being:

1. Participants involved in performing work measurement usually have other jobs and responsibilities. Therefore, Standard Work tends to become something performed on a part-time basis at best, which leaves little time to become proficient at the task or to be driven to make it a full success.

2. It has been thoroughly proven that most operators under observation usually perform at a pace and often a method that doesn't represent normal conditions. There are numerous reasons why this happens. However, the point is that sound work measurement practices take both method and effort into respect and adjusts performed time to more adequately represent a normal pace.

Conclusion

Given a basic understanding of an automobile, almost anyone can change a spark plug. But delving into the inner workings of a combustible engine and doing a dependable job requires an experienced mechanic. It is absolutely no different when it comes to the science of work measurement. Standard Work will be much more reliable for making crucial management decisions pertaining to manpower requirements, cost estimates, and the like if the job is performed by a qualified Industrial Engineer. In addition, the more advanced aspects of work measurement (the re-methodization and re-redistribution of labor) can be much more professionally and effectively applied.

Fact #2

In the vast majority of undertakings, Poke-Yoke and SMED are usually last on the list of priorities. If and when they are performed on equipment, the work is seldom done by someone who is fully qualified, trained, and formally dedicated to the project.

Associated Misconception

The theory is that when improvements are scheduled in specified areas of the factory, the assigned team members will decide if and when such work is important; they will then make the necessary change or see that appropriate talent is brought in to assist as required.

The Problem

Once again, there are some serious flaws with the theory. To name a couple:

1. Unless a qualified engineer has been assigned as a formal member of the team, the likelihood of the matter ever coming up would be small. Great attention is usually applied to the more basic aspects of Lean. Therefore, little time if any is left for more sophisticated applications, such as applying SMED and Poke-Yoke.

2. Assuming that the first problem noted is somehow addressed and fully resolved (which would be highly unusual), little actual improvement would be made.

The reason is doing the job effectively requires devoting attention to *all* key equipment. In the same respect mentioned for Standard Work, effectively doing the job of SMED and Poke-Yoke requires someone with a high level of skill and expertise in equipment engineering, who can be devoted to the effort full time.

Conclusion

Being effective with SMED and Poke-Yoke requires an engineer or group of engineers who are assigned the task for a specified period of time, until the work is fully completed. Anything less is not going to bring a plant's equipment up to World Class standards. Putting aside Ohno's frequent speech about teamwork and better utilization of the production worker, I've seen enough to believe the more critical tasks associated with the Toyota Production System were performed by highly qualified engineers — what Ohno refers to as "Toyota Style Industrial Engineering."

This type of front end work has been minimized in most Lean Manufacturing efforts undertaken on U.S. soil. Instead, crucial tasks such as Work Measurement, Mistake Proofing, and Setup Reduction have been turned over to participants who lack the skill to perform the job effectively.

Ohno tended to group separate types of engineering under one category called *Toyota Style Industrial Engineering*. He noted,

IE is the use of techniques and systems to improve the method of manufacturing. In scope it ranges from work simplification to large-scale capital investment plans. IE

has two meanings. One aims at improving work methods in a plant or in a particular work activity. The other one means the specialized study of time and action.

In the United States, where by his own admission Industrial Engineering was born, the work he specifies is typically split between two fields of applied engineering. One is Manufacturing Engineering, the other Industrial Engineering. Although a large portion of the skills and educational focus are similar, the applied expertise at a plant level is usually substantially different.

As working experience goes, the manufacturing engineer tends to be more mechanically inclined, the industrial engineer more oriented toward methods and procedures. I make this distinction because it is important to understand that using only one of the engineering skills noted can sometimes leave an operation short of needed ability. As a past manager of Industrial Engineering, I carry a high level of respect for both functions. But for anyone to think that a typically experienced mechanical engineer can effectively do an industrial engineer's work (or vice versa) is misinformed. There is a unique difference in assigned focus and ability, and a definite need for both.

In *Fast Track,* I addressed the importance of Industrial Engineering under the topic *Your Greatest Warriors Against Waste.* However, I now believe the specific application of both Industrial and Manufacturing Engineering is required on the front end of a Lean initiative in order to adequately move the process in the right direction. The distinction between the two is important because the initial step outlined for *Level One*

Lean (LL-1) rests with getting *all* key production equipment in the plant up to a prescribed level of performance.

The question will arise, I'm certain, as to precisely how many engineers would be required for such an effort and how long it should take. The required number of engineers would vary depending on the complexity of the equipment involved and the size of the operation. But the goal should be to see that the initial work outlined doesn't exceed a year and preferably less, if at all possible.

As a first step, the assigned engineers should receive advanced training in the principles and concepts of Lean Manufacturing, with special focus on SMED, Standard Work, Poke-Yoke, TPM, and Workplace Organization. The assignment of two full-time engineers, ideally one mechanical and one industrial, to begin the process of appropriately engineering equipment would be essential.

THE CONTINUOUS IMPROVEMENT FACTOR

I've already elaborated somewhat on this, but in the simplest way of putting it, we have to somehow change management expectations and the focus of the participants in the process, from making certain we get it right the first time to making certain we kill batch manufacturing and do it as fast as humanly possible.

This was the focus of many of the early proponents of Lean, such as Richard J. Schonberger. In his book *Building a Chain of Customers*, he reflects, "*The real problem is commitment, and that's the purpose of this book and others like it, plus the many other forms of spreading the message. I can only say the ball is in your court. Do your utmost to educate any in the company who will listen, get little fires lit (small proj-ects started) everywhere, and don't stop for sign-offs at every step."*

In building on the principles of *Waste-Free Manufac-turing*, I strongly encouraged the same approach, but took it a step further by noting, "Just Do It — But with knowledge." The knowledge that's most important is focusing early atten-tion on getting a plant's key production equipment up to an acceptable level of compliance in order to provide a stable foundation for continuous improvement.

Key Reflections

- We've come to overlook the influence a plant's key equipment, which was designed to support a batch manufacturing environment, can have on speed of implementation. The equipment can often slow an initiative to near creeping progress.

- One of the more serious flaws in the way Lean is approached in the United States is the fact that there is no universal means of approaching the task and measuring progress. Once a firm gets past a basic commitment to Lean, the focus has to be placed on keeping the process moving forward rapidly. Just as important is making certain the tools of Lean are used to their ultimate level of benefit.

- The techniques used by the Japanese consultants were essentially "bundle implemented" over a very short period of time. This was necessary in order to demonstrate precisely how the tools were intended to work. On the other hand, these same tools and techniques were developed and deployed over four decades at Toyota. Therefore, we should not take for granted that the methods the early Japanese consultants used represent how we should go about installing the process.

- Because of what Lean Manufacturing has come to mean in the United States, more attention has been focused on techniques aimed at making small improvement rather than on a well thought out strategy aimed at making a factory waste free, as quickly as humanly possible.

- Many in our industry are trying to position themselves where Toyota currently stands after forty years of effort. The only difference is Toyota is working to improve processes that have been highly engineered (in SMED and Poke-Yoke) and, for the most part, we aren't.
- We have to realize there are fundamental steps that should be taken and be wise enough to see they're fully accomplished. This calls for recognizing the rules of engagement, and embracing the steps required to implement the process in a timely and orderly fashion.

2

AS A PLANT'S EQUIPMENT GOES,
SO GOES LEAN

The basic assumption behind the typical approach used to launch Lean Manufacturing centers on:

1. Demonstrating the power of cross-functional participation in upgrading shop floor processes
2. Setting the stage for a new way of doing business
3. Sending the message that management is willing to put time, money, and effort into the process, and that it wants everyone's input in making lasting improvement a lasting reality.

However, this approach almost always leaves a plant dealing with backup and supporting equipment that isn't capable of complying with good Lean practices. If equipment is truly in tune with what a firm is striving to accomplish with Lean, progress will go relatively smoothly. If not, equipment can and often becomes a drain on the process. Therefore, to a large extent, the way a plant's production equipment goes, so goes Lean!

I would point out that the connection between equipment that isn't adequately geared to support Lean and a basic lack of commitment from production leadership generally goes hand in hand. This combination can literally paralyze a Lean initiative. While the prescribed pilot area may be striving to run in a pull fashion, the rest of the factory is still typically working with a push system of production. As a result, supervision continues to focus on things they've been used to doing and by which they've been measured and otherwise programmed to value, often for years on end.

Middle managers and, more specifically, those assigned to running the day-to-day activities of production must be clearly inspired to support Lean. The best way, of course, is to tie job performance to compensation. But a very important element boils down to making their job easier, by removing their single most serious stumbling block, which is equipment that isn't capable of adequately supporting the challenge.

Typical flaws with the insertion of Lean tend to say to the workforce — through an inability to abide by the most fundamental principles — that the company really isn't all that serious about the process. To some degree I believe those of us working in the field are partially to blame for this. We've suffered from a paradigm that boils down to believing the most important thing is to train as many employees as possible in Lean and turn them loose to make change on the shop floor. This paradigm is the very thing we preach to others that they should avoid. Yet, we're done that ourselves and it simply isn't getting the job accomplished. So what's the answer?

The answer is changing the course of action, but doing so when we fully understand where the focus should be directed. The matter of applied equipment engineering has been seriously downplayed in the United States, principally due to the perception that came out of the early Japanese influence on the process. But I have *repeatedly* seen the process falter after the initial stage of implementation. The major reason is that proper expertise was not applied to the sciences of SMED, Standard Work, and Poke-Yoke.

What generally occurs is production equipment in the factory continues producing in the same manner it always has. Newly-established inventory levels become theoretical at best. As far as the plant itself is concerned, little if any actual improvement ends up being made. Therefore, it would seem we should at least examine the possibility of how to go about inserting Lean in a more effective manner — one that would shift emphasis on precisely where the process should start and proceed from there.

A Lean initiative, intent on making the transition as fast and effective as possible, would take a three-fold approach. The first would be focused on getting the plant's key production equipment up to speed, followed by the education and motivation of the general workforce, and last changing the product flow through the insertion of Lean principles plant wide. Approaching the initiative in this manner would make for a far less chaotic and smoother transition than what is typically achieved with the more common method of implementation.

As a first step, the assigned engineers should receive

advanced training in the principles of Lean Manufacturing, with a special focus on SMED, Standard Work, Poke-Yoke, TPM, and Workplace Organization (see Figure 2.1). Assigning two full-time engineers — ideally one M.E. and one I.E. — to begin the process of appropriately engineering equipment is essential.

The key to the exact number of engineers needed to complete this most important first step depends largely on how quickly an operation desires to see Lean fully accomplished. In most cases, it would not take an army of engineers. But even if a few engineers are required, a company will find it's actually a smart investment — one which will almost immediately begin to pay dividends.

As far as the breakdown of tasks performed by the engineers, the first objective of the assigned M.E. should be to implement SMED on key production equipment in the factory. The I.E. assumes responsibility for applying the tools and techniques of Standard Work on the same processes. Once both tasks are fully completed, the M.E. shifts attention to mistake proofing (Poke-Yoke), whereas the I.E. focuses on Workplace Organization (WPO). As a last step, they team to implement TPM on the equipment (see Figure 2.2).

Assuming that a plant was staffed appropriately with the needed engineering support, key production and maintenance supervision personnel should be given a special Lean Manufacturing Orientation just prior to starting work on the shop floor. The purpose would be to help them understand the basics and, more importantly, their precise role in helping to see the initiative accomplished. (see Figure 2.3).

Figure 2.1 Window Diagram — Training Outline

	Day 1:	Day 2:	Day 3:	Day 4:	Day 5:
WEEK ONE:					
TRAINING IN SMED & STANDARD WORK	*INTRO TO JIT MFG	*INTRO TO SMED	*ADV. SMED	*ADV. STD WORK	*WRITTEN EXAM
	*THE HIDDEN WASTES	*INTRO TO STD. WORK			*DISCOVERY
	*FLOOR EXERCISE	*FLOOR EXERCISE	*FLOOR EXERCISE	*FLOOR EXERCISE	*SENIOR MGT. PRESENTATION
WEEK TWO	DAY 1:	DAY 2:	DAY 3:	DAY 4:	DAY 5:
TRAINING IN POKE-YOKE & TPM	*INTRO TO MISTAKE PROOFING	*POKE-YOKE	ADV. POKE-YOKE	ADV. TPM	WRITTEN EXAM
	*INTRO TO TPM	*TPM			*DISCOVERY
	*FLOOR EXERCISE	*FLOOR EXERCISE	*FLOOR EXERCISE	*FLOOR EXERCISE	*MGT. PRESENTATION

As noted, the training consists of both classroom and floor work, progressing through the week to more advanced training in the subject matter and wrapping up with a written exam and a discovery process (with appropriate feedback), before ending with a presentation to senior management.

Figure 2.2 M.E. and I.E. Focus

STEP	M.E. FOCUS:	I.E. FOCUS:
ONE ▶	SMED ▼ ▼	STD. WORK ▼
TWO ▶	POKE- YOKE ▼	W.P.O. ▼ ▼
THREE ▶	TPM	TPM

Figure 2.3 Window Diagram — Lean Orientation

DAY 1	DAY 2
PLT'S LEAN STRATEGY INTRO TO LEAN MFG. DISCOVERING THE HIDDEN WASTES FLOOR EXERCISE WRAP-UP	UNDERSTANDING THE VALUE OF SMED AND POKE-YOKE UNDERSTANDING YOUR ROLE IN THE PROCESS GROUP PLANNING SESSION WRAP-UP

We must look realistically at the matter of implementing Lean Manufacturing. How can we expect a plant superintendent, a production foreman, a maintenance head, and any other vital support personnel to actively work against what has been drilled into them for years, without first providing a full understanding? The answer, of course, is we can't. If we somehow fool ourselves into believing otherwise, we've simply headed down the wrong path.

Accomplishing Lean Manufacturing requires a serious commitment on the part of management — one strong enough and willing to totally revamp the goals and objectives that have driven the company for years. (On the other hand, we assume management would not be looking seriously at implementing Lean Manufacturing, if it didn't already feel the need was evident.)

Figure 2.4 indicates the training focus at each stage of progress. As shown, the first detailed training is provided to the assigned engineering staff, followed with awareness training for production and maintenance supervision. At Level

TRAINING LEVELS:

PROGRESSION AND FOCUS FOR MOST
EFFECTIVE TRAINING APPROACH

FUNCTIONS

LEVEL THREE
SUPPLIERS
OTHER BUSINESS

LEVEL TWO
PRODUCTION WORKFORCE
DIRECT OFFICE SUPPORT FUNCTIONS
MAINTENANCE PERSONNEL

LEVEL ONE
ASSIGNED ENGINEERS
PRODUCTION SUPERVISORS
MAINTENANCE SUPERVISORS

Figure 2.4

Two, the training emphasis is shifted to the general work-force, including hourly, salaried, and as many supporting functions as possible. In Level Three, training encompasses suppliers and other business functions. I will speak more to this topic in Chapter Five, under *Performing Adequate Training*.

MEASURING PROGRESS AND TRACKING RESULTS

Adequately measuring progress has been difficult when it comes to Lean. This area is something I feel very strong about and managed to keep in mind when I was searching for a better approach. Therefore, under the guide-lines established for fully achieving the various levels below, measurement becomes a much simpler task.

Figure 2.5 addresses four levels of implementation. Within each level are a number of specific items that must be accomplished. Each level carries a weight factor based on the depth of work required (see Figure 2.5). Here, the precise levels are established; the specific accomplishments required are few in number (no more than four), making tracking and communicating accomplishment of milestones much easier.

It is important to clarify that once a plant has reached a Level IV, it doesn't mean work is done. In practice, the work required to make a plant Lean and keep it that way is never truly finished. Toyota is still at it, after forty plus years. On the other hand, once a plant has achieved a Level IV, it is well on its way to making a noted shift from implementing the process to fully sustaining it.

LEVEL FOUR LEAN (GOAL: 18 TO 36 MONTHS
1. OVER 80 IN-HOUSE EVENTS CONDUCTED - Wgt. Factor 30%
2. MINIMUM OF 50% OF ALL KEY SUPPLIERS CERTIFIED Wgt, factor 30%
3. SOME FORM OF LEAN INCENTIVE BUILT IN Wgt. Factor 20%
4. LEAN CONCEPTS DRIVEN INTO OFFICE/ADMIN. PROCESSES Wgt. Factor 20%

LEVEL THREE LEAN (GOAL: 16-18 MONTHS
1. PULL SYSTEM OF PRODUCTION FULLY IMPLEMENTED THRU-OUT SHOP- Wgt. Factor 40%
2. 100% OF ALL KEY EQUIPMENT HAS ERROR PROOFING APPLIED- Wgt. Factor 40%
3. MIN OF 60 IN-HOUSE IMPLEMENTATION EVENTS CONDUCTED - Wgt. Factor 20%

LEVEL TWO LEAN (10-16 MONTHS)**
1. A SELECTED "LEAN PILOT AREA" ESTABLISHED AND FULLY REVISED – Wgt. Factor 40%
2. LEAN AWARENESS ORIENTATION (ALL NEW/EXISTING EMPLOYEES) – Wgt. Factor 10%
3. MIN OF 20 IN-HOUSE TRAINING AND IMPLEMENTATION EVENTS CONDUCTED – Wgt. Factor 10%
4. MIN OF 50% OF ALL KEY EQUIPMENT HAS ERROR PROOFING APPLIED – Wgt, Factor 40%

LEVEL ONE LEAN (GOAL: 10-12 MONTHS)
1. SMED/TPM/STANDARD WORK APPLIED TO ALL KEY EQUIPMENT - Wgt . Factor 50%
2. FULL TIME LEAN COORDINATOR ASSIGNED - Wgt. Factor 30%
3. LEAN AWARENESS TRAINING (ALL PRODUCTION/MAINT. SUPERVISION) - Wgt. Factor 10%
4. 18-4 MONTH LEAN IMPLEMENTATION PLAN COMPLETED AND APPROVED - Wgt. Factor 10%

**WHERE OVERLAP TO NEXT LEVEL CAN BEGIN

Figure 2.5 Effectively Tracking Progress

SUMMARY OF LEVEL ONE LEAN

- All key production equipment have SMED, Standard Work, TPM and WPO thoroughly applied.
- A full-time Lean Coordinator is hired or promoted from within.
- Lean Awareness training for the plant's maintenance and middle management production functions are fully accomplished. Written goals and objectives for these individuals are revised accordingly.
- An on-going training program is established for new employees entering the ranks outlined.

As you can see, the first step is focused on getting a plant's key production equipment up to speed. In doing so, an overlap of work can begin at Level Two, as time and resources are made available, *after* a full-time Lean Coordinator has been appointed. However, it is considered best to wait on overlapping milestones until an 18–24 month implementation plan has been completed and approved by senior management. This plan will be addressed in more detail in Chapter Five.

SUMMARY OF TRAINING PHASES AND SPECIFIC CONCENTRATION

Figure 2.6 summarizes the various phases I am proposing and the specific type and concentration of training to be accomplished in each.

PHASE:	TRAINING FOCUS	TRAINING LEVEL	PERSONNEL INVOLVED
I	SMED POKE YOKE WPO STANDARD WORK	ADVANCED *	ASSIGNED ENGINEERS
	LEAN AWARENESS	BASIC **	PRODUCTION MANAGERS MAINTENANCE MANAGERS SELECT HRLY & SALARIED
II	LEAN SPECIFC	EXTENDED ***	ASSIGNED HOURLY AND SALARIED TEAMS
III	FULL INCORP OF LEAN	EXTENDED	MAJOR WORKFORCE

SPECIAL NOTES:

*Advanced: 3–4 weeks of intensive, in-depth training in Lean, with a special focus on SMED, Poke-Yoke, Workplace Organization, and Standard Work.

**Basic: 1–4 hour general overview pointing out the flaws and wastes associated with Batch Mfg and the considerable benefits of pursing a Lean Mfg initiative.

***Extended: On-going 3–5 day training and implementation events aimed at bringing a plant up to a Level One Lean status and setting the stage for higher levels of achievement, which will be covered in more detail as we move forward.

Figure 2.6

WHAT SPECIFICALLY EARMARKS A PIECE OF EQUIP-MENT AS "KEY"

Production equipment that is used on a variety of products should generally be considered key, as this designation pertains to equipment that has been engineered to:

1. Change from one product or run to the next in single minutes (less than ten minutes)
2. Eliminate recurring mistakes in processing (common scrap and rework problems) pertinent to the equipment in question.

Note: If it is deemed that single-minute changeover or error proofing cannot be achieved on a key piece of equipment without spending capital, the assigned engineers would notify the Lean Manufacturing Coordinator, who then decides if they should do the best they can and move on or if a special capital appropriation is warranted.

The first and most fundamental step called for in approaching these tasks is to select an appropriate team of people to investigate and categorize every piece of production equipment in the factory. Once done, the team then prioritizes the list in terms of the depth of work required to bring it up to a proper level of compliance, from easiest to hardest.

This ranking most often rests with how complex a particular piece of production equipment is. It is important to note that most factories have equipment which should not be viewed as key, especially equipment that requires little-to-no setup or adjustments and which poses no substantial issue when it comes to downtime, scrap, and rework. To meet the specific definition of Key, the following applies:

- Any and all equipment that takes more than single minutes to setup (nine minutes or less)
- Any and all equipment that has a history of scrap, rework, or excessive downtime
- Any and all equipment failing to meet 100% of its stated capacity, in terms of throughput, quality, or uptime.
- Any and all equipment that is historically unsafe or unreliable

The equipment listed as the toughest challenge should be addressed first. Some would argue it would best to start with something easier and less complicated and, in doing so, establish a successful example. I sincerely believe this is the wrong approach. The sooner that the more troublesome processes in a factory are addressed and brought up to minimum Lean compliance standards, the better an operation is going to be overall.

KEY MANAGEMENT COMMITMENT AT LEVEL I

Each of the four levels covered point to one or more key commitments that must be made by management in order to be totally successful. In the case of Level I Lean, the first commitment is to recruit and hire a full-time Lean Manufacturing Coordinator for the factory, one who preferably has experience in leading a Lean initiative. The second commitment is to provide the Coordinator with a full-time staff, consisting of at least one Manufacturing Engineer and one Industrial Engineer. The basic qualifications for the I.E. and M.E. are covered in more detail in Chapter Five.

In situations where I've proposed hiring a full-time Lean Coordinator and two engineers for a start, I've met far more resistance than cooperation. Even when I've gone on to point out if the job is done right, they can pay for themselves tenfold and more, I often have to struggle to get management to buy in. But if a manufacturing operation isn't willing to make such an investment in its future — an investment which is actually small in relationship to the gigantic benefits that can be derived — they should probably rethink starting a Lean

Initiative, because it is likely doomed for failure. They can chip around the edges and make some improvements, of course, but in all probability, they will never truly be a qualified Lean Manufacturer.

SUMMARY OVERVIEW OF LEVEL II LEAN

After achieving Level I Lean, there can be argument as to what should go into accomplishing Level II and beyond. Certainly discussion is warranted as to the precise number of levels involved. I believe almost everyone would agree there is no such thing as an ultimate level of achievement, and there certainly is no finish line in striving for continuous improvement. But I would remind the reader that achieving Level I means that SMED, Workplace Organization, TPM and Standard Work have been applied thoroughly to *all* key production equipment in the factory. In addition, supporting maintenance personnel and production supervision have been successfully aligned to support the effort.

Level II Lean boils down to aggressively driving the process forward. Progress is made through extensive training of the workforce, inserting the fundamentals area by area, and establishing a pull production environment. In addition, steady refinements to the work done in Level I are made as required (i.e., further attention to setup reduction and mistake proofing, as the need requires).

Accomplishment of Level II starts with the selection of a Pilot Area., This area is used as a showcase, representing what the entire factory will eventually look like over time. The pilot project is best accomplished through a comprehensive, two-week training and implementation event involving some of the best people in the factory. Focus is placed on making major flow revisions, along with implementing some solid workplace organization and the application of Standard Work. Success measurements should center on floor space reductions, overall staffing requirements (in terms of total employees needed to perform the work), improvement in quality indices (scrap, rework, defect rates, etc.), and manufacturing lead time reductions (from point of order to finished product, ready to ship).

Using what was learned with the pilot project, work then begins in conducting as many "fast track events" as possible. These events drive the process forward, through the entire factory, steadily increasing overall training and awareness. Details about the two-week comprehensive lean event and the three-day fast track event" can found in Chapter Four. A plant has to conduct a minimum of twenty fast track events in order to satisfy the standard for Level II. Therefore, even with a goal of two events per month, which is relatively aggressive, this effort would require up to a full year to complete this and, in most cases, 14–16 months realistically.

A KEY MANAGEMENT COMMITMENT AT LEVEL II

In order to be successful in accomplishing Level II Lean, management must give Lean as much attention and respect as it does any other area of the business. For example, the Lean Manufacturing Coordinator should report directly to the plant manager and be an active member of his or her staff, attending all staff functions and participating in staff planning sessions. In addition, a reasonable operating budget should be established for the Lean Office; this budget should include a special account for maintenance work to rearrange equipment and facilities in support of the changes required on the shop floor. Anything less will not take the factory up to a level II status.

SUMMARY OVERVIEW OF LEVEL III LEAN

Assuming an aggressive schedule, the timetable noted for full accomplishment of Levels I and II suggests a plant has been into Lean for approximately eighteen months. At this point, the factory, its management, and its employees are either comfortable with Lean and positive about where it is taking the operation or they have otherwise given up. It is also at this point where the more advanced aspects of Lean start to come into play. Special consideration must be given to a number of important issues, specifically including:

1. What will it take to tie factory schedules directly to customer orders? How can a Pull system of production be fully instated plant-wide?
2. Which administrative processes should be subject to Office Kaizen or Value Stream Mapping, in order to better support a Lean environment?
3. Who are the top 15–20 suppliers and how will they be encouraged to involve themselves in the process?

KEY MANAGEMENT COMMITMENT AT LEVEL III

In order to be successful in achieving Level III, management must commit to involving administrative functions such as

Accounting, Materials, and Scheduling actively in the Lean process. This involvement goes up to and includes performing office Kaizen or Value Stream Mapping in order to fine tune a function so that it fully supports a pull production system for the factory, which is absolutely essential to taking the next step up the ladder.

SUMMARY OVERVIEW OF LEVEL IV LEAN

Achieving Level IV status positions a plant so that it can start to take full advantage of Lean, in terms of profit improvement, an exceptional level of flexibility to customer demand, and future growth in volume. It also positions an operation so that it starts to take the matter of continuous improvement seriously by considering:

1. Which manufacturing processes, if any, should be looked at as outsourcing candidates?
2. How should the factory be rearranged to take full advantage of the space savings gained, and more effectively support the Lean process?
3. What other opportunities exist to further improve key equipment and facilities?
4. What are the company's training and communication activities going to be in the future in order to fully support a more advanced level of continuous improvement?

A KEY MANAGEMENT COMMITMENT FOR LEVEL IV

To achieve a Level IV, management must commit to look at the operation from the standpoint of what it does best and where it should place its resources. This includes the potential of outsourcing some processing and importing others, along with gaining appropriate approval as needed for funding a major re-layout of the factory.

Adding it up, a plant on a relatively aggressive schedule could see achievement of Level IV in two years. However, three years would be closer to reality. Three years may be viewed as a long time, but we should be striving to be as close in parity as we can with companies such as Toyota, which has been at the process for decades. From that perspective, three years to achieve a Level IV status could be seen as a rather astonishing feat. Picture how the factory will both look and perform at that point:

APPEARANCE

The factory will have the appearance of being extremely organized, exceptionally clean, and thoroughly efficient. Working aisles will be free of clutter and have a highly polished look about them. Everything will have a specified place, marked accordingly, with everything in place. Inventory will be minimized to the point that it appears almost non existent. Fork lifts will give way to an established number of transferable bins that are used as needed between work centers.

Any and all work-in-progress and purchase component storage will be located at point of use. Visual controls will abound. Information regarding customer orders and performance to established "Takt Time" will be posted in every work center, along with information about how the area is doing in terms of downtime, quality, scrap, rework, and such.

Production workers will be busy, but not performing at a frantic or frustrating pace. The bottom line: the factory will look and act like a well-tuned machine, front to back.

PERFORMANCE

The factory will be meeting customer requirements better than ever before — in fact, substantially better. Setup and changeover will be dramatically minimized whereas manufacturing flexibility will be extremely high. Scrap, rework, and obsolescence will be almost entirely eliminated. Any production issues will be addressed immediately and fully resolved. Workers will understand the exact throughput required, down to the second, if needed. If and when an infrequent difficulty arises, they will automatically stay after normal hours to see that daily schedules are fully achieved. Noticeably fewer operators will be needed to meet the same and very often increased levels of output. But overtime will be far less frequent. Workers, on average, will be much more satisfied and comfortable with their assigned tasks. Operating budgets will be met reliably and consistently because surprises and inconsistency in performance will be significantly reduced.

At this point, someone might reflect that I seem to be implying manufacturers should simply trust the process enough to pursue an activity such as SMED and Poke-Yoke to its fullest, no questions asked. The answer to that is absolutely yes! Although there is indeed a place for thought and planning, it should only be in relationship to what it takes to get the job done, as quickly as possible. Reducing setup and eliminating equipment errors on the front end of Lean initiative is the right thing to do. The same is true for both professionally and adequately applying work measurement and TPM on all key equipment in a factory. So simply do it — and waste no time getting there!

KEY REFLECTIONS

- The connection between equipment that isn't adequately geared to accommodate Lean and a lack of production leadership support generally go hand in hand — and can literally paralyze a Lean initiative.

- If a factory isn't willing to make the kind of investment noted, which is small in relationship to the gigantic benefits that can be derived, it should probably rethink getting into Lean.

- The matter of applied engineering has been seriously downplayed in the United States, principally due to the perception that came out of the early Japanese influence on the process.

- Putting aside for the moment Ohno's frequent speech about teamwork and much better utilization of the production worker, the more critical tasks associated with the Toyota Production System were performed by highly qualified engineers and, more specifically, what Ohno refers to as "Toyota Style Industrial Engineering."

- Not every piece of equipment in the factory will be viewed as key. This especially applies to equipment which requires little to no setup and adjustments in order to change from one product to the next, and which pose no substantial issue when it comes to the matter of downtime, scrap, and rework.

- Reducing setup and eliminating equipment errors on the front end of a Lean initiative is the right thing to do. This is also true for professional and adequate work measurement and applying TPM to all key equipment in the factory. So simply do it — and waste no time getting there!

3

ACCEPTING THE NEED FOR CHANGE

ASSESSING THE FACTORY

In understanding clearly the need for change, it's appropriate to know where you are if you hope to know where you're going and — most important — whether you're fully prepared to make the journey (see Figure 3.1). To know where you are, you should make a simple, yet powerful assessment of a factory.

The Primary Objective

The goal is not just to implement such tools, but to know when and how to use them

Figure 3.1

The individual recommended to lead this exercise seemingly should be the factory's plant manager. But this role isn't always limited to this particular position. The most important criteria rest with the individual who sets the vision for the factory. That person is ultimately responsible for day-to-day decisions in terms of everything from labor requirements to meeting assigned budgets and forecasts.

My experience shows that the plant manager isn't always the proper individual; the definition and role of plant manager varies greatly from organization to organization. Whoever is the right person, however, should tour the factory, accompanied by a number of key staff, and explore the four most noteworthy signs about how the operation is doing. These signs are:

1. Inventory
2. Storage
3. General Flow
4. Visual Controls

Inventory

One of the clearest signs about how far an operation has advanced toward becoming World Class is to observe the amount of inventory on the factory floor. Anything more than the amount required to service the job being performed is simply too much inventory. Inventory in the form of parts and raw material resting outside of an imaginary six-foot circle of the equipment involved is probably inventory that isn't being immediately utilized. Its presence is usually a warning sign that the factory has not, as yet, put proper emphasis on such programs as SMED, Kanban, and Poke-Yoke.

Storage

Storage racks should be a well-identified home for needed tools, dies, fixtures, etc. If racks are being used for any other purposes, you likely have a problem waiting to happen, and certainly an environment that isn't in tune with good Lean practices. Piece part inventory of any kind should *never* be stored in a rack unless a clear quality, safety, or environmental reason warrants it. The general manager of an Irish manufacturing facility once said to me, in response to my observation that there wasn't any storage racks in his factory, *"Well, as I see it, John, the only good racks are the ones our competitors are using!"*

Look for power transfer conveyors. These can appear to be extremely efficient, but they are notorious for holding huge batches of work-in-process inventory. They are a telling sign that the factory has not yet effectively addressed or accepted the need for point-of-use-manufacturing. The same is true of pace conveyors used to transfer work from one operation to the next, with the expressed intent of pacing operator performance and output. They may look efficient, but they clearly minimize flexibility and eat up valuable floor space.

General Flow

Try to get a reading on the general flow. Does it appear continuous in nature or does it wander aimlessly from one area of production to the next, sometimes even retracing itself? Look for stoppage points, where parts and components are stored until they can reenter the flow again at some point. Don't confuse or accept intermittent storage as an obvious Kanban, because most often that will not be the case.

Also look for "production monuments" in the flow. For example, large presses may require (or are perceived to require) formed pits in the floor. But pits make it difficult, if not next to impossible, to move and rearrange presses as needed. Check if the flow is essentially departmentalized (e.g., a welding area, a press area, a sub-assembly area). Departmentalization is usually one of the best signs that the factory is largely stymied from a flexibility and inventory control standpoint.

Visual Controls

Check the extent that working aisles are utilized. If the only clearly marked aisles are for material movement (most often referred to as the plant's main aisles and used for fork lift traffic, etc.), there are not proper floor controls in place to visually aid flow. Check to see if there is a clearly identified place for everything and that everything is in its place. Even a trashcan should have a properly identified location. Also check for the use of "shadow boards" that act as a home for tools, which insures that tools are being properly maintained and accounted. Review operator instructions to see how readily available they are to the operator and if they are being properly updated and utilized. Check how many visual examples are utilized. In well-tuned operations, 50% or more of operator instructions are visual as opposed to written.

PERFORMING THE MORE INTENSE FOCUS GROUP ASSESSMENT

It is always best to start at the receiving door and work forward. Before starting, various individual assignments should be made. One team member should keep count of every time the flow of parts and components stop in order to be stored and accounted for. This information alone can be a real eye opener in most batch-oriented operations.

Another participant should keep a running record of how many places in the operation have a clearly-marked system of problem identification and resolution (for example a flip chart or whiteboard established for the purpose). In most cases, this type of feedback will not be found, but it's good to make the point.

In addition, another participant should be asked to record every time operators are seen doing something other than such value-added work as drilling holes, brazing assemblies, painting components, and assembling units. In a batch-oriented environment, it isn't unusual to see 30–40% of the operators doing non-value-added work such as searching for a tool to complete the job, waiting on setup, sorting parts, reworking units, and the like.

The group should thoroughly examine the following eight questions:

1. Is inventory (raw material and work-in-progress) sitting idle, not being immediately being utilized?

2. How long is needed to go from one setup or production run to the next?
3. Is there full assurance that operators have what they need to do the job (the proper materials, tools and equipment)?
4. Are all production processes free of scrap and rework?
5. Is preventive maintenance performed on a daily basis? Are the operators personally involved?
6. Are machines completely free of dirt, oil, and grime? Have equipment leaks been entirely eliminated?
7. Does every section of the factory know how it stands in terms of meeting schedule and maintaining quality? Do they know the specific skill levels of each and every employee?
8. Do all production areas work in unison toward a common objective? More specifically, do supervisors and operators recognize a clear link between what they are currently producing and actual customer orders?

These questions, individually and collectively, provide the framework needed for evaluation in conjunction with the seven key warning signs addressed in my earlier book *Fast Track to Waste Free Manufacturing*.

INVENTORY SITTING IDLE

If you do nothing else in your daily treks through the factory, examine how much inventory you have sitting around not being immediately utilized. This information will help keep you on track with the task at hand and how much progress is actually being achieved.

One of the most crucial messages that should be communicated to employees — especially shop floor employees and production supervisors — is that the plant *must* reduce its inventory levels dramatically. This reduction may even go as far as creating some part shortages and, thus, some added overtime if necessary to stress the point about reducing inventory.

Before anyone becomes skeptical about adding potential overtime in order to encourage lower inventory levels, allow me to point out that nothing good is normally accomplished without some degree of sacrifice. In this case, I'm referring to the possibility of adding some overtime in order to become better at what you do. The cost of added overtime should actually be viewed as an investment in helping a plant to stay on track to becoming World Class. But to clarify my point, let me share this personal experience:

I was asked by a high level executive if I would take a tour through one of his factories with the appointed plant manager and provide him some feedback. As the tour progressed and I went about making various observations, the plant manager asked how I could simply stride down the aisle and make "flat assumptions about his operation." He added that I actually knew little, if anything, about the processing required to make the products he produced.

My response was that I based much of my judgment on the enormous amount of inventory he had floating around the factory. When he asked me to clarify my remark, I told him that one can look at inventory much like a doctor searching for the source of an ailment. If more inventory than is needed is there, it's a symptom of a problem. If it's there in mass, the problem is likely to be deadly serious. The most important point that I stressed to him was that carrying enough inventory can hide problems that are dearly costing the operation (see Figure 3.2).

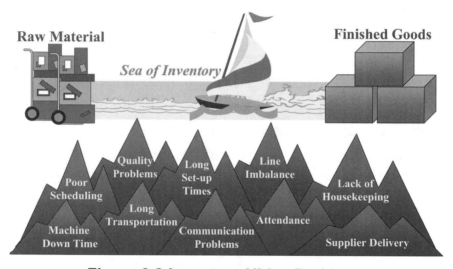

Figure 3.2 Inventory Hides Problems

Most batch manufacturing operations of any size carry inventory levels that run hundreds of thousands of dollars. This means a fixed level of investment has to be semi-frozen in order for the factory to open its doors to doing business. The total will vary, of course, depending on the product produced, the time required for suppliers to fulfill commitments, and the level of customer orders.

In my experience, companies that enter into Lean

Manufacturing find they are capable of cutting established inventory levels by as much as 80%, and sometimes more. But they seldom do. The question then becomes why and the answer rests in two factors.

The first factor is that most batch manufacturing plants have operated with a "just-in-case" mentality for years. Just in case the supplier doesn't deliver. Just in case the machine breaks down. Just in case Joe gets sick and doesn't show up for work because he is the only one in the factory who really knows how to run the equipment. The list goes on.

The second and more puzzling factor is management's all-too-frequent refusal to get out of the way of progress. Rather than clearly supporting the effort and trusting a well-proven process, management fears the unknown and second guesses the actions taken. When management balks at making change — and it's been my experience it happens more than most would like to admit — what sort of message do you suppose they send the troops?

If you carry enough inventory, then wastes and ineffi-ciencies in the factory can be covered up. If fully exposed, the cost is often staggering. If you lower the inventory level, the real problems will begin to surface, which is precisely what should occur. Then you can identify where work should be applied to insure the problems are fully resolved and never surface again (see Figure 3.3).

The bottom line: when it comes to the matter of inven-tory, both management and employees must accept that *just enough* is precisely where the focus should be. No more, no less! The one external exception deals with supplier delivery.

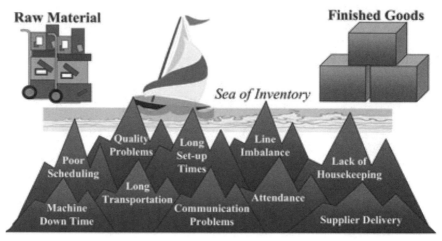

Figure 3.3
Lower the inventory level to find/resolve problems

In some cases, a plant may indeed be forced to carry more than enough inventory until it can establish a reliable vendor certification program, This program weeds out those suppliers who either cannot or will not thoughtfully partner in the march toward becoming a Waste-Free manufacturer.

LENGTHY SETUP AND CHANGEOVER

I've always been amazed by the lack of adequate attention most factories place on setup and changeover. I have seen operations where extensive efforts were aimed at measuring everything from the work performed by direct labor employees down to the smallest element of time. In those same factories, however, little to absolutely no measurement was made of those responsible for seeing that setup or changeover was accomplished quickly and efficiently.

Changeover is the length of time from the last good piece of one run to the first good piece of the next run.

last good part   next good part

Time between is emphasized because it is a period of non-value added. In other words, the equipment is idle and not producing parts.

Figure 3.4 Definition of Changeover

Before going further on this subject, let's consider the difference between setup and changeover. **Setup** is the work associated with physically removing and inserting dies and making needed machine adjustments. In other words, setup is the time spent between the last good part, from one run, to the first good part from the next run.

Changeover pertains to physically removing materials and components, then replacing them in order to ready a process for production (see Figure 3.4). For example, a changeover is needed when moving from one product to another on a final assembly line.

Once an order has been completed, the process often requires the help of a trained setup person to prepare for the next order. This setup person may or may not be readily available, and may not feel compelled to be as quick as possible

about completing the job. Recognizing and correcting inept-
ness within direct labor usually ranks high on the priority list
of most manufacturers But inefficiency in the indirect labor
ranks can often run rampant and never be seriously challenged.

The problem boils down to the fact that there are clear
standards for the work required on most direct labor jobs, but
usually few standards, if any, for indirect labor work. I'd be the
first to admit there usually isn't much direction given to indirect labor work, one way or the other. The scenario often goes
like this:

*Joe was a great machine operator who took it upon
himself to learn how to set up his own equipment. He proved
to be exceptionally good at it. So good, in fact, a decision was
made to make him a floating setup person. In that role, Joe
was asked to make certain that all the machines in the
department were set up properly. He went on to learn much
before eventually moving on bigger and better things. Joe
wasn't asked and therefore didn't bother to document how he
went about performing setup. As other people moved in and
out of the job, they ended up taking their own particular
approach. Time went by, and setups began taking longer than
ever. Furthermore, the quality of setups began to deteriorate.
In the end, a decision was made to refer setup back to the
individual operators, who were then trained how to do the job
— ironically by the same people who had proven to be totally inept at it.*

In most manufacturing firms, a flexibility problem usually centers on the factory's key production equipment. Upon

taking a close look, you will usually find:
1. Little if any planning is done in advance of a pending setup.
2. Much time is spent searching for tooling, fixtures, materials, etc.
3. There is no standard method prescribed for conducting the setup.
4. There are far less-than-adequate checks and balances in place to assure the setup is performed properly.

Setup and changeover that hasn't had a high level of SMED applied usually has a tremendous amount of waste. This waste can be corrected once you understand that setup is composed of two distinct components: Internal and External. The internal component deals with what needs to take place while the machine is down (idle). The external component deals with what can and should be done while the machine is up (running.)

A sizable portion of the time required for setup can often be shifted from Internal to External. For example, dies, fixtures, and material can be pre-staged prior to the time the last piece of the current run has been completed. In addition, you can usually find ways to speed up setup by using quick disconnects (clamps rather than nuts and bolts). Standardizing die heights avoids the need for machine adjustments to accommodate variations (see Figure 3.5).

If setups were truly looked at as waste, it would be much easier to do what's necessary to take corrective action. Unfortunately, many operations wrongfully see setup as value

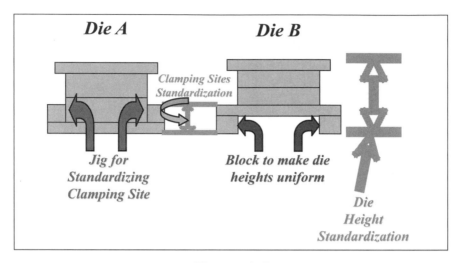

Figure 3.5
Standardization of Die Heights and Clamping Sites

added, which isn't the case. Perhaps one of the best ways of showing why is to think in terms of the customer.

Customers would ideally prefer to see a manufacturing firm press a button and instantly spit out the product they need. Because this isn't feasible, they accept (or value) the fact that parts have to be produced and assembled in order to achieve fit, form, and function. On the other hand, customers place absolutely zero value on how much a manufacturing process is required to push parts around the factory or on how long it takes to change from one product to another — nor should they!

Manufacturers will sometimes argue that many of their indirect employees require special skills, over and above those required to run a simple piece of equipment. This may be true. However, it is only so because management has chosen to make their direct labor operators little more than robots who load machines and push buttons. Instead, management should train them to do much of the work required to keep their equipment functioning properly, while making certain that the parts produced are of the highest quality.

For example, take away the setup person and the inspector checking quality. Train and turn those responsibilities over to the machine operators. You will have fewer people doing a much better job of meeting customer requirements. You'll also have more than enough justification to pay the people who are adding *real* value more for the work they do.

If a football team has a poor rushing defense, the coach can address the problem by putting his entire team on the line to guard against the run. However, this leaves it totally vulnerable to the pass. Either way, the weakness results in touchdowns by the opposition. Unlike football, where the score against the competition is always clear, many manufacturers tend to keep score on themselves. In doing so, they make misleading assumptions about how well they are actually performing. Their evaluation is usually based on a set of prescribed measurements, aimed at pumping out more and more inventory in order to cover any and every problem. Meanwhile, they are blind to the wastes that are allowing the opposition to score repeatedly, virtually unchallenged.

Toyota set an initial goal of performing setups in something less than ten minutes — thus the development of SMED (Single Minute Exchange of Dies). However, improving setups to less than *half* of what they currently are is a good start in the right direction (see Figure 3.6).

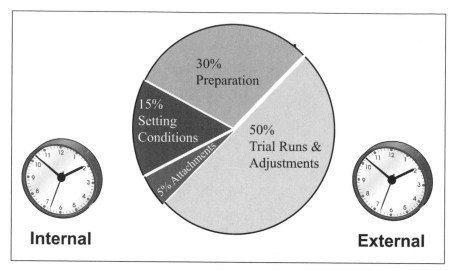

Figure 3.6 Breakdown of Changeover Time

LACK OF PROPER TOOLS AND MATERIALS TO DO THE JOB

One of the major problems with most batch manufacturing is the time lost by operators searching for what they need to do the job. Most often this search involves simple items such as hand tools and materials. In truth, the time wasted has little to do with whether an operation is batch or not. But it has been my experience that Lean-minded factories tend to pay far more attention to this particular issue.

To a large degree, the issue boils down to attitude. Although managers may complain that an operator had to spend time looking for a tool, they seldom take constructive steps to resolve the problem for the future, such as the use of tool boards (see Figure 3.7). Changing attitude is a vital focus for those using the guiding principles of Lean Manufacturing, because even small amounts of time, multiplied by the mass-

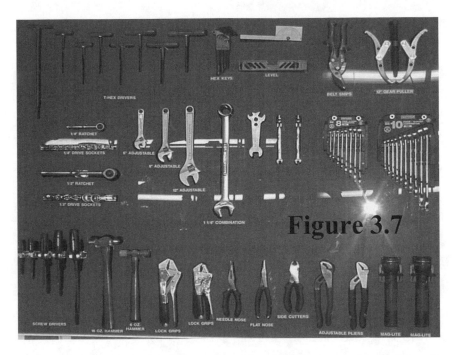

Figure 3.7

es, often result in high levels of inefficiency and lost time for an operation.

HYDRAULIC AND OTHER TYPES OF EQUIPMENT LEAKS

Equipment that has oil, hydraulic, or other types of leaks should be a flag that preventive maintenance procedures are not being adequately performed. Many batch manufacturing firms have a "run until it breaks" mentality. For them, preventive maintenance is a once a year thing; it is generally performed during an annual two-week shutdown period for employee vacations. But why do you suppose so much emphasis is placed by world class manufacturers on a spotless shop floor? To impress someone? Allow me to assure you that isn't the case.

First, a very clean and tidy environment looks better

and makes for a nicer place to work. More important, the driving objective is to provide the ability to instantly detect if equipment is potentially developing problems. When you walk down an aisle and see oil under a piece of equipment, do you have any idea how long it has been there or if the contributing problem needs attention? Probably not. However, when you see the same problem in a factory that keeps its equipment and shop floor spotless, what do you suppose it tells them? Correct! That something is about to go wrong.

Many aspects of the Lean process are so fundamentally simple that people have a difficult time accepting how important they really are. Solutions often seem far too simple for helping to resolve equipment problems and other difficulties facing the typical batch manufacturer. Unfortunately, manufacturers have become conditioned to believe that answers will always lie in new and expensive capital investments. They forget entirely that they may be overlooking ways to substantially increase the output and reliability of the equipment they already have — and often the improvements are very dramatic!

In most manufacturing operations, if the same level of energy given to new investments was instead applied to improving the capability and output of existing equipment, we would see some remarkable differences on the shop floor, and for the better. I've heard too many people comment about the difficulties associated with the "tired old equipment" in their factory, when the actual problem was far from that. Equipment may indeed be old and there may indeed be a time to replace some of it. Yet it's usually "tired" because it suffers from inadequate upkeep.

Taiichi Ohno of Toyota said, *"If a machine purchased last year has been poorly maintained and produces only half its operable rate, we should regard its value as having declined 50 percent."*

We have to think again about Toyota in its infancy. After World War II, it was only one of a number of industries in Japan struggling to rebuild the economy. At the time, Toyota could in no way compete with America in terms of huge capital investments. It therefore had to find ways to steadily improve the capability of its existing equipment and facilities. In doing so, it developed SMED, Poke-Yoke, and the bedrock fundamentals for what later came to be termed Total Productive Maintenance (TPM).

Toyota wasn't interested at the time in having a show-case facility, where one could eat off the floor, so to speak. It kept its shop spotless in order to know beyond a doubt if and when a piece of equipment was developing problems. At the first sign of a leak of any kind, Toyota was extremely quick to address and resolve the problem; always striving to keep in mind a solution that would make certain the problem didn't happen again.

FLOORS SATURATED WITH DIRT, OIL, AND GRIME

This topic and the preceding one go hand in hand; thus, some further clarification is warranted. Seriously cleaning and polishing the shop floor is sometimes seen as nothing more than show. Although this reason certainly isn't the case, it is why production operators and supervisors often take the process less than seriously. They haven't been given the needed understanding. They don't see a distinct advantage in every aspect of Lean — putting on a show doesn't come into play.

A production foreman once told me he didn't care much for Lean. He could appreciate management wanting a spotless shop to impress visitors, but "spit shining" it took valuable resources away from getting product out. In one respect, he was right. If all one is concerned about is pumping out parts and components in order to meet a prescribed schedule (most likely based on an inaccurate forecast), then anything that seriously distracts from production will be viewed as a waste of time. What he and others fail to recognize is that, in order for the firm to remain competitive, both the operating rules and the overall focus of manufacturing has to change.

For argument sake, let's assume there are two factories — one batch and one Lean — that are making the same products. They are able to put out the same level of throughput over the course of a given month. What then justifies Lean over batch? First and foremost, Lean producers recognize that the game isn't satisfying customer needs at any cost. Almost any manufacturing firm can do that, given the freedom to wade neck deep in waste. The batch manufacturer will always require *more* overtime, *more* scrap, *more* rework, *more* obsolescence, *more* on-hand inventory, *more* space, and *more* people to do the same job — always!

Therefore, striving for floors that are free of clutter, dirt, and grime isn't just a game to improve appearance. It is a *must* in order for a firm to become truly World Class, in both its thinking and shop floor practices.

CLUTTERED WORKING AISLES

There's hardly a factory that hasn't gone to the trouble of marking off its aisles. But many completely ignore the fact that aisles are there for a specific purpose. That purpose is traffic — to keep people, equipment, products, and supplies on the move. When I've asked various managers why things were stored inside established aisles, the most frequent response I get is, "Well, it's not supposed to be that way." My immediate reaction is to ask, "If that's the case, then why hasn't someone done something about it?" The answer usually boils down to a lack of discipline and focus. To be fair, however, I will agree batch manufacturing isn't totally without focus. The problem is that the focus is usually directed at producing huge volumes of parts and inventory, whether they're absolutely needed or not.

Special efforts have to be made to insure that employees are aware of the proper use of aisles. Under no circumstances should aisles have material stored inside them, even for the briefest period of time. To illustrate the importance of aisles, I'll relate this personal experience:

I was consulting with a firm that had been in business for well over four decades. We were in the process of implementing Lean in a pilot area on the shop floor and were making some decent progress. The firm was expecting some very important potential customers and I was invited to participate in a discussion about preparations for the visit. I learned that the customer had been in the plant before and that they weren't impressed with what they saw. The owner was doubtful the previous impression was going to change any. When the opportunity arose, I brought up the fact that a difference in perception could be achieved by simply shaping up the finished goods area and clearing aisles throughout the factory. To make a long story short, management took the advice. I was later informed that the visitors were not only favorably impressed, but went about placing an immediate order.

Note: Nice clean floors and well marked and uncluttered aisles leave a potential customer with a favorable impression, even if it is obvious to the trained eye that far too much inventory is still sitting around (see Figure 3.8).

I'm sure other factors came to bear on the customer's decision. But it does show the kind of positive influence a

Figure 3.8

neatly-arranged facility can have on customer perception. Other reasons for unobstructed aisles pertain to safety and cost. If a pallet of material is left unattended in the aisle and employees move from behind it into the aisle, they could be stepping into the path of an on-coming fork truck or maintenance vehicle. Even more likely, the pallet itself could be struck, thus damaging the contents and setting up the need for scrap or rework.

LACK OF REAL-TIME SHOP FLOOR INFORMATION

One certainly doesn't have to be a Lean manufacturer to know how a production area is performing in terms of meeting customer demand, downtime, scrap, rework, etc. But I've yet to see a batch-oriented factory where one can walk into any given area and readily acquire such information.

The reason this happens in most batch manufacturing operations is there is generally little, if any, concern about how a given production area is doing in relationship to the rest of the plant. Employees are measured, disciplined, and rewarded on the basis of how much product is being pushed through the shop. Generally they could care less if their work has any immediate tie to firm customer orders. They are taught that machines are going to break down, that material won't always be available as needed, and that keeping direct labor efficiency up is much more important than keeping inventory down.

If I sound like I'm being unduly critical, that truly isn't my objective. In most cases, we're speaking about very good people with extremely good intentions; who are normally highly energetic about doing what they believe is expected of them. What I'm trying to point out is the depth of change in thinking that is needed in manufacturing.

Knowing how every production area is doing in terms of what is required to meet firm customer requirements is essential for any company aspiring to a World Class manufacturing status. In the Lean shop you will normally find control boards at every work center. These provide everything from an operator skills matrix to a posted list of problems and concerns, along with other information designed to help department operate more efficiently.

The major difference between the batch manufacturer and the one using Lean is that if the batch manufacturer chooses to deal with such issues, it is usually addressed the following week or month, in some prescribed production meeting designed for the purpose. The Lean manufacturer will address and resolve problems *immediately*, on a real time basis, without exception.

USE OF OVERTIME TO MEET NORMAL PRODUCTION SCHEDULES

When overtime is consistently required in order to meet standard production schedules that a factory is theoretically manned and equipped to achieve, there are unquestionably some very serious underlying problems. Most often, the excuses boil down to the following:

1. A machine or piece of equipment broke down.

2. Various components or finished goods were placed on hold.
3. There was a shortage of parts and raw materials.
4. Someone was missing who carried the skills needed to do the job.

These are only some of the dozens of excuses made, but I'm certain you get the picture. Such reasoning is inherent to batch manufacturing. They are due in part to some of the factors already mentioned, along with others yet to be discussed. Best put, an operation should applaud the identification of problems and abhor the use of excuses. A problem should be viewed as something that has been brought to attention so it can be resolved and doesn't happen again. An excuse, on the other hand, is nothing more than a problem gone unresolved or a solution gone unused.

On more than one occasion, I've made the assertion that most excuses can be entirely eliminated. I've been challenged by shop floor management, who insisted I wasn't speaking about the real world. In their judgment, machines *were* going to break down; there *were* going to be unavoidable shortages of parts and components, and finished goods *were* going to be placed on hold — because it was simply the nature of the beast.

This resistance points to the influence paradigms have on performance. People who have lived for years in the world of machine breakdowns, parts shortages, and product holds aren't going to be swayed easily into believing there's a way to free themselves of the restrictions that have plagued them in the past and created endless hours of added work and effort. Add to the list the overtime used to meet standard levels of throughput, which is something you will commonly find in batch manufacturing.

In the latter stages of advancing Lean, which will be discussed in the final chapter, there are ways to go about building added flexibility in both the workforce and equipment. Thus, overtime for favorable reasons, such as taking on added business, can be reduced over time. But this normally involves creating an atmosphere of flexibility and utilizing concepts such as Mixed Model Scheduling.

UNTIDY AND DISORGANIZED WORK AREAS

The foundation for continuous improvement is Workplace Organization. This concept was identified as one of the four principles of Waste Free manufacturing. Work areas in the factory that are untidy and disorganized are always a clear warning sign of the need to change.

With good workplace organization, one will find a work area that is:

1. Free of trash, dirt, oil, and grime
2. Well defined, with regard to a place for everything and everything in its place
3. Free of unneeded tools, fixtures, equipment, and sur plus inventory
4. Seasoned with visual controls
5. Highly efficient and orderly, with flexibility built in

If a trash can is needed in a work area, a specific place should be marked for it. The same holds true for machinery and inventory. Everything has a designated place and is expected to be there, if it isn't being immediately utilized. This may sound a bit juvenile, but it has a very important purpose. You should know, without a doubt, if the process is working as intended or if something is out of order.

Speaking of *order*, there is not a word that better and more specifically defines Lean Manufacturing. Order is the oar that steers the entire Lean ship. Without it, the conduct of manufacturing is extremely chaotic and undisciplined — something easy to see during a quick walk through of the factory. If you walk by an area and see a shadow board with a screwdriver missing, but none of the workers are using a screwdriver at their immediate work station, what does it tell you? That the screwdriver is missing, of course. But think about the common manufacturing work area that hasn't applied a Lean focus. What do you think the chance would be of even knowing the screwdriver was missing in the first place? Little to none.

I often refer to Workplace Organization as providing the ability to truly see the process. It's like taking off a blindfold that has long hindered us from understanding precisely how the process was intended to work. Before good Workplace Organization is performed, the work area is usually cluttered with things that added absolutely no value and, in fact, place restrictions on getting the job done.

KEY REFLECTIONS

- In understanding clearly the need for change, it's appropriate to know where you are if you hope to know where you're going and, most important, whether you're fully prepared to make the journey.

- If setups were truly looked upon as waste, it would be much easier to do what is necessary to take corrective action. Unfortunately many manufacturing operations wrongfully see setup as value added.

- People comment about the difficulties associated with "tired old equipment" in their factory when the actual problem is far from that. Equipment may indeed be old and there may indeed come a time to replace it, but it's usually "tired" because it suffers from inadequate upkeep.

- Order is the oar that steers the entire ship. Without order, the operation of manufacturing is an extremely chaotic and undisciplined beast — something easy to see during a quick walk through of the factory.

- The Lean way of doing business centers on addressing and resolving problems *immediately* and doing this on every occasion, *without exception*.

4

SETTING THE STAGE

Doing a decent job of implementing Lean requires an examination of what I would describe as the "peripheral issues" at hand. These have nothing to do with the process itself, but what needs to take place in establishing an appropriate atmosphere for change.

MOTIVATING EMPLOYEES TO DO WHAT COMES UNNATURALLY

Every firm that is thinking about making Lean a full success should consider the matter of motivating and encouraging its workforce to do things in a way that will *not* feel natural to them, at least in the beginning. For those who have run a batch manufacturing operation for years, this unnatural feeling is probably the single biggest hurtle they will face. Putting aside the typical opposition to Lean, with the more crucial opposition often coming from middle management rather than operators on the shop floor, we have to face the fact that it is only human to resist something that doesn't feel right.

If a long lost tribe were suddenly discovered in the far recesses of some deep, dark jungle, and were given rifles, many would toss aside their new weapon at the first chance and eagerly go back to using their spears. Why? Because something, which directly affects what people have been used to doing for an extensive period of time, creates a sense of suspicion — even if they're shown evidence of its superiority.

When it comes to Lean Manufacturing, the concerns of the workers can be extensive and far ranging. These concerns include: Will the change eliminate my job? Will it require me to do something I'm not prepared or capable of doing? Is it a ploy aimed at laying off employees? In all likelihood, telling employees that Lean is a better way of doing things will not do the trick. They will generally put little stock in that alone. "Better" has been scrutinized and resisted throughout the ages.

I don't profess to be an expert when it comes to precisely how to go about instituting the proper motivation. The truth is there's no magic bullet that applies to every situation. But it is something that needs to be thoroughly addressed and it has to go beyond the old proverbial pat on the back. A couple of considerations include:

1. A firm guarantee that no employee who is displaced as a result of implementing Lean will lose employment. Instead, they will be assigned roles that serve to support the process, such as fill-in to allow training for others in Lean. Saying this doesn't mean there will not be adjustments in the workforce based on common business factors, such as a decline in sales or volume.

However, it does send a message that the company will not penalize workers who are affected by the implementation of Lean Manufacturing.

2. The development of a Lean incentive plan that includes measurements which clearly serve to promote the process. Such incentives can include a bonus for specific Lean accomplishments, such as fully achieving one of the various levels of Lean. Another incentive can be a merit plan which earns points based on specified accomplishments, directed at an employee educational assistance program.

There are numerous other programs and policies that can be examined. But the key is come away with something that has some real meat in it and serves to get employee's attention. If they are executed correctly, these programs can more than pay for themselves. I'd like to say such measures aren't necessary, but I'd be seriously remiss if I did. These measures are indeed needed in most manufacturing operations in order to advance Lean in a reasonably meaningful fashion.

GETTING MIDDLE MANAGEMENT ON BOARD

In his book *The Leader — A New Face for American Management*, Michael Maccoby describes four types of managers found in large corporations (see Figure 4.1). Maccoby writes, *"No type is superior to another. As a leader, each has positive and negative aspects that make him or her effective in some situations, ineffective in others."*

TYPE (and key word)	POSITIVE TRAITS	NEGATIVE TRAITS
Craftsman (quality)	*independent, exacting*	*uncooperative, inflexible*
Jungle Fighter (power)	*brave, protective*	*ruthless, dominating*
Company Man (service)	*loyal, prudent, caring*	*servile, fearful, soft*
Gamesman (competition)	*daring-risk taking*	*rash-gambling*
	manipulative-seductive	*dramatic-inspiring*
	fair-flexible	*unfeeling-unprincipled*

Figure 4.1 Maccoby's Four Types of Managers

It's been my experience that middle management can be one of the more serious stumbling blocks to a Lean undertaking, if not the most significant barrier. It ultimately doesn't really matter which category of manager they tend to fit because each style has its own pluses and minuses. What matters is whether a particular style will prove to have a positive or negative impact on the effort.

I do not profess to share Maccoby's vision. However, I can address a group of management styles in a slightly different manner — and I have absolutely no problem providing some firm opinions about these!

The MY WAY OR THE HIGHWAY Managers

We've all seen them. The managers who have their mind made up about everything. They aren't particularly interested in new ideas because anything that tends to distract from the status quo simply doesn't fit in their mind. This type of manager has to be the first to go if an organization hopes to make Lean Manufacturing a success. This is because Lean is a process that counts heavily on cross-functional interaction and, most important, the active thoughts and ideas of everyone. This manager doesn't understand or respect teamwork and, therefore, can be a very serious liability.

The I JUST DO MY JOB Managers

In fairness, these managers are usually reasonably good at what they do, having focused their entire careers on it. In addition, if given a specific task, they will generally do their best to comply. But these are also individuals who don't like change. Not because they're against it; they've never really been against anything their entire career. It's because they greatly fear they may not possess what it takes to comply. The end result is that they will usually go along, but only marginally support the effort.

The MY PEOPLE COME FIRST Managers

These are the managers who think their chief obligation is to defend their subordinate's work. They take any criticism of their department personally. They can often be among the more difficult when it comes to something new and different. They are generally well thought of by their people and less than respected by others, who tend to see protectionism at work, very often at someone else's expense.

The LET'S GIVE IT A TRY Managers

These managers are usually well respected and extremely knowledgeable of the plant, its products, and processes. They are most often good at getting their point across to others and are hesitant to ask people to do something they're not willing to try themselves. They typically have the company's best interest in mind and never shrink from a challenge, especially if it has the potential of serving the betterment of the organization. This type of manager is the kind every good Lean initiative needs.

Ideally, what you are looking for among the middle management ranks are those managers who — even if they give the impression that they may have their own misgivings about the process — possess the ability to support the process, without prejudging its merits. If and when the perception is anything less, it is probably time to dig deeper and take appropriate action

THE PLANT'S LEAN TEAM

This team consists of the people who are most actively involved in driving the process forward. I've seen the group called many things (e.g., The Lean Coordination Team, The Lean Fast Track Team). What's important is that they're recognized as such and given an appropriate status that not only includes a chief responsibility to the process, but the authority that should go along with it.

As plant manager, I elevated the person heading this effort to a staff level position. This step provided the person with as much clout and decision-making authority as any other member of my staff. When I held morning business meetings, the attention and reporting on Lean was just as in-depth as anything else on the agenda — and, to be totally honest, higher in some respects. But if that was a fault, I don't apologize.

On the other hand, I fully recognize that many opera-tion managers would not be comfortable going as far as I did, especially those who have little first hand experience with Lean. Therefore, the reporting level of the Lean Coordinator will vary greatly from one operation to the next. Regardless, attention has to given to insuring that this particular position carries an adequate amount of clout and authority to get the job done.

Ideally, the plant manager holds a bi-weekly meeting with the Lean Coordinator and his or her staff. The meeting is designed to address progress made against an 18-Month Implementation Plan. Holding a weekly meeting is admirable, but may be a bit too often to track and report results effective-ly. The key to success, however, is to keep a high level of atten-tion on progress. Stay focused on the objectives spelled out in the plan. Make time to discuss where problems may have developed and what steps should be taken to keep the process on track. Frankly, anything else is a waste of valuable time.

PICKING APPROPRIATE LEADERSHIP FOR THE PROCESS

Picking the right person for the role of Lean Coordinator is extremely important. It may even be the single most important selection a plant manager has to make regarding Lean. Given the right coordinator, the process will proceed almost flawlessly. But selecting the wrong person for the job can prove to be a very serious stumbling block to success. Avoid thinking that any conscientious employee who has reasonable communication skills and is given proper training would suffice. This simply isn't the case. The individual selected has to have some very special qualities, which include:

- An open mind, along with the understanding of the need for change
- Preferably two-to-three years experience in manufacturing, with a first hand knowledge in the tools, concepts, and techniques of Lean Manufacturing
- An excellent communicator, capable of getting thoughts over to people in a highly reliable fashion
- Someone who genuinely likes people and understands the power of cross-functional participation in problem resolution, but who can be depended on to hold strong convictions when it comes to maintaining discipline about the approach

OPEN MINDED AND DEDICATED

The individual toward whom people will be looking to provide training, guidance, and direction has to be opened minded and dedicated to change. There simply isn't room for even the slightest skepticism or doubt on the part of this leader. They are, in essence, the conscience of the company when it comes to Lean.

ASSOCIATED WORK EXPERIENCE

The individual should have some practical experience with Lean, preferably in moving a factory from a batch manufacturing approach to a pull system of production. For this reason, the appropriate person is not often someone who is promoted from within. Whoever is chosen, however, should put aside past learning to some extent and bring great energy to leading a new and different approach to the business of production.

ABILITY TO COMMUNCATE EFFECTIVELY

The individual should possess proven communication skills. If there is one thing I've noticed in the selection of people to lead the Lean process, it is a lack of attention to this important element. Candidates interviewing for the position should be able to indicate clearly how their communicative skills have influenced others in a positive and results-oriented manner.

STRONG PERSONAL CONVICTIONS

The individual should express a genuine desire to work with people in both addressing and resolving issues. In turn, they should express the ability to communicate their thoughts and ideas to others in a manner that isn't perceived as threatening or argumentative. At the same time, they have to avoid sacrificing, in any way, the fundamental guiding principles and the day-to-day disciplines that are key to the process.

THE IMPORTANCE OF DEDICATION TO THE PROCESS

One of the most important characteristics of good Lean Manufacturing Coordinators is a devoted dedication to the process. They should be willing to stand up and fight hard for the process as needed. This doesn't mean they need to be single minded or less than understanding about other company initiatives and undertakings.

There's much truth behind the old saying that "Manufacturing is just one damn thing after another." Manufacturing can certainly be a very chaotic and frustrating place. To a very large degree, it stands alone between product engineering and marketing on one side and the customer on the other. It is fully expected to deliver in a manner that best represents the way the product was designed, in terms of quality, function, and reliability; yet it is also required to do so in a timely and effective manner — excuses be damned!

That, of course, is precisely why so much emphasis has been placed over the past decade in advancing the sizable advantages of Lean. But in the real world, it is hard to find anyone outside of manufacturing who seriously worries if standard lead-times have been slashed in order to meet a sales promotion or to guarantee a special promise made to an important customer. Nor is it easy to find someone outside of manufacturing who is overly concerned about flaws in design that fail to meet good design-for-manufacturability guidelines. I could go on, but the point is that a certain amount of frustration is built into manufacturing, which isn't always self induced. As a result, a special undertaking such as Lean requires an earnest level of "stick-to-it," in order to see the process through to completion.

Considering the fact that Lean is touted as being a never-ending process can sometimes make the matter all the more difficult. Therefore, an essential part of any Lean Manufacturing Coordinator's job is seeing that the process remains on track. Doing so will require standing thoughtful yet firm if and when things begin to slip. The importance of staying the course under the worst of conditions was well put by a close associate of mine, who once remarked, "Lean is a long and hard journey that takes only a few steps down the wrong track to find yourself facing an oncoming train."

Keeping the operation on the right track and headed in the right direction is the primary job of the Lean Coordinator. It's good to recognize that this focus usually requires a special kind of individual. However, finding such a person is far from impossible if the proper effort is applied. Unfortunately, there have been way too many instances where good, solid contributors were assigned the role, only to discover they didn't carry the enthusiasm or skills to carry the program to completion.

In my own experience selecting the Lean Coordinator, I decided after great deliberation to ask one of my key staff to assume the role. The person happened to be someone who was highly respected by the rank. The major problem I faced in convincing him to take the job was that he managed one of the largest departments in the factory, with over forty people reporting to him. I was asking him to give up a high-level managerial position in order to take the role of an individual contributor. You can imagine his first reaction. But he eventually agreed to take the job and has since gone on to become a highly respected consultant in the field of Lean Manufacturing. Very satisfying to me was the fact that he went out of his way afterwards to assure me he was highly indebted for steering him in the right direction.

I'm not at all insinuating the Lean Coordinator needs to be someone of the level I appointed. However, I do feel much more attention needs to be placed on this important matter than I've typically seen. Once selected, the coordinator and assigned staff should be located on the shop floor, with a sign over the door saying "Lean Coordination Office." Ideally, the office should have an open door policy, where any employee is welcome to step inside and bring up a point, ask a question, or make a suggestion in support of Lean.

Clear and open communications are essential to any successful Lean Manufacturing effort. The job of Lean Coordinator requires someone who is knowledgeable enough about a plant's production processing to speak intelligently about it. All in all, it takes somewhat of a juggling act to be totally effective in the job. But there is some helpful advice I can give to anyone assuming the position: Let the established principles guide the effort. Although there is indeed room for give and take on the part of coordinators, they should never allow themselves to sacrifice any of the guiding principles, such as those spelled out in *Fast Track to Waste-Free Manufacturing* (Figure 4.2):

To give a better example of what I'm driving at, assume you are the Lean Coordinator. The production manager requests that you delay applying SMED for a couple of weeks

Principle	Associated Tools for Implementation
1. Workplace Organization	5-S, 6-C, Andon, Std Work
2. Uninterrupted Flow	1 Piece Flow, Point of Use, Std Work
3. Insignificant Changeover	SMED, Work Place Design, Std Work
4. Error Free Processing	Poke-Yoke, Std Work

Figure 4.2
Guiding Principles of Waste-Free Manufacturing

on a particular piece of equipment because (right or wrong) he believes it would be far too disruptive to the current schedule. As coordinator, you can safely agree to this request. But if the production manager says he'd like a particular piece of equipment to be totally exempt from the process, you should try to impress upon him the utter importance of insuring that all production equipment is brought up to minimum Lean standards. If the production manager remains adamant, however, you are left with little choice than to raise the issue with senior management and press for a compromise that would best meet everyone's agenda — including the one undertaken for Lean Manufacturing.

KEY MANAGEMENT POSITIONS THAT STRONGLY INFLUENCE SUCCESS

When it comes to implementing Lean Manufacturing in a fast and efficient manner, you must have the full support of some of the factory's high profile managers, including:

- Production Manager
- Maintenance Manager
- Engineering Manager

Ideally, an organization is continually seeking to build as many of the traits of excellent managers as it can into its existing staff. However, time isn't always on your side when it comes to Lean. Making a complete and thorough transition will take a factory up to 36 months — and that's with an aggressive and positive group of middle managers. In essence, what it boils down to is: Do you have the time and inclination to force feed an attitude adjustment in your existing middle management ranks before the process is solidly underway?

I cannot stress this next point strongly enough. Assessing the ranks of the key middle manager ranks is essential before leaping into Lean Manufacturing, only to find out later something that should have been addressed on the front end. In most undertakings, it's appropriate to explain the mission and provide managers with a chance to prove their support and enthusiasm. However, Lean doesn't happen to be one of them. The reason is most initiatives simply do not cover the scope of change that Lean Manufacturing requires — not even close! Here, we're speaking about changing *everything* about how we go about the business of manufacturing. Non-believers who hold key management roles can literally paralyze the effort.

Nothing frustrates me as much as key manufacturing managers who profess to support a Lean effort, then fundamentally proceed to work against it. Their reasons for doing so could be many. These reasons could range from the feeling that Lean creates an unneeded distraction to having an honest feeling it really isn't all that important. But the managers should consider the fact that Lean Manufacturing isn't some passing fancy. At the least, something all good managers should consider is to make certain they *don't get in the way of those who are working to make it a success.*

Good commanders know their people have to work on more than one front in order to win the war. In the case of Lean, we're speaking about a war both on waste and against the competition. The job isn't going to get easier as time goes by. My plea then to the key non-believers would be to give the process a chance to work and, where you can, lend a reasonable amount of positive support and influence.

THE PLANT MANAGER

Without the full commitment and support of the plant manager, the incorporation of Lean is almost doomed from the start. In all fairness, however, the best plant manager in the world cannot make a difference if senior management, who effectively control the purse strings, doesn't fully support the effort.

One might ask why an operation would choose to pursue Lean if its senior management wasn't fully convinced it was the thing to do. Unfortunately I have seen more than one example of this; if and when it happens, no amount of personnel assessment and reorganization at a plant level is going to change things.

Under normal circumstances, however, the plant manager ends up being the key person who can literally make or break the process. Most plant managers are highly dedicated, energetic, and foresighted. The fact remains, however, that many plant managers are highly self serving and glued to the old way of doing things.

I've personally seen the top brass of a company entirely fooled about their plant managers — I've witnessed the face that some plant managers put on in the presence of senior management vs. quite another in performance of their job. So, what's the answer? Fire the existing plant manager and hire someone who has proven experience in Lean Manufacturing? That isn't practical, of course. But what *is* practical? Insure that the plant managers very clearly understand the company's intentions of making the process an absolute success. Let them know —with absolutely no doubt — they will be measured by appropriate benchmarks and rewarded on the merits of their personal involvement.

I've come to recognize something about plant managers who are highly successful in the role. It isn't what they know that makes them successful, but how they react to conditions that aren't of their own making. The highly successful plant managers find a way to balance their convictions with those that come to bear on the industry. They proceed to work as effectively as possible in supporting the effort. If you happen to be a plant manager who isn't a firm believer in Lean, or otherwise see it as somewhat of a stumbling block to your own plans and convictions, you can nevertheless make an effort to strike the balance noted. I can assure you, you won't regret.

THE PRODUCTION MANAGER

Assuming senior management support is in place and

the existing plant manager is a Lean advocate, the next position that needs to be assessed carefully is the Production Manager. This position goes under many names and titles. Here the focus is on any individuals who have responsibility for overseeing the day-to-day activities of the factory. These managers must hold a sincere appreciation for what Lean Manufacturing can bring to the operation. They should be personally involved in supporting change on the shop floor, whether this essentially boils down to a hands-on role or simply lending strong, on-going support to the effort.

THE MAINTENANCE MANAGER

Maintenance plays an enormously important role when it comes to implementing Lean Manufacturing. Simply put, without a highly-focused maintenance team, a Lean Initiative has little chance of success. Unfortunately, some maintenance managers are so steeply embedded in the status quo they're extremely reluctant to accept a new and revised role for the maintenance function. This attitude will have to change. They will be asked to give up some of the maintenance duties they have been entirely responsible for in past or in the least share those responsibilities with others.

One of the most critical facets of change for maintenance requires transferring much of the obligation for machine upkeep to the machine operator. However, this provides more time for Maintenance to play a lead role in the development of a sound Total Productive Maintenance (TPM) process throughout the entire factory. In addition, Maintenance holds an important obligation for actively expediting on-going changes on the shop floor that serve to enhance uninterrupted flow, workplace organization, insignificant changeover, and error free processing. How energetic the maintenance function is in meeting these challenges will have a lot to say about how successful a Lean initiative will be.

THE MANUFACTURING AND INDUSTRIAL ENGINEERING MANAGER

As previously discussed, these important functions have been far less than effectively utilized in Lean initiative in the United States. In addressing those operations that have separate Manufacturing and Industrial Engineering functions, all managers should be well versed in the benefits of Lean and be supportive of the process. They should also be willing to give up one-to-two engineers (depending on the size of the organization); these engineers would ideally report to the Lean Coordinator for a period of 12–18 months. During this time, key production equipment in the factory would be brought up to minimum Lean standards.

These functions simply have not been used to their fullest potential. Collectively, they should be the stalwarts of the process. Unfortunately, this hasn't happened on a grand scale in the United States.

Following the guidance from this book will help you involve both functions much more actively in the process. Organizations that find a way to make Lean Manufacturing the primary mission of both their industrial and manufacturing engineers will see some truly remarkable results. With minimal added resources, this can be accomplished without sac-

rificing the regular on-going duties that each function normally carries (e.g., bills of material, operation sheets, line balanc-
ing, work measurement).

Back in 1989, I made a trip to Japan to visit a number
of manufacturing firms. At Mitsubishi I found a troop of
Industrial Engineers assigned exclusively to progressing their
version of the Toyota Production System. Even in much
smaller Japanese operations, Industrial Engineers were
actively involved in implementing and refining the process.
Since then, I've visited and worked with numerous firms in the
United States and have found nothing that even remotely
compares.

Although it is difficult to draw any firm conclusions, it's
safe to say there's room for a much more active role in
America's Lean undertaking by Industrial and Manufacturing
Engineers. For this to happen, however, much depends on
how willing the leadership of these particular functions are to
step forward and on how the company's top management
establishes and provides the opportunity.

SEEING THE PROCESS IS TAKEN SERIOUSLY:

As addressed in the Introduction and again in Chapter
One, there is a problem with employees seeing no end to the
process of Lean Manufacturing. Implementation is hard and
can, in fact, be extremely exhausting. Therefore, one of the
more serious issues with most undertakings is finding a way
to keep the process actively moving forward. In most cases,
finding the way means taking basic human behavior into

respect, not withstanding the influence of modern day technology.

In *Organizational Behavior and Personnel Psychology*, authors Kenneth N. Wexley and Gary A. Yukl address various characteristics of the workforce that pertain to basic human behavior. Their work makes for excellent reading and is almost timeless with respect to how and why people react to various situations in the workplace.

But a contemporary factor that has to be taken into account is the instant feedback from media, the Internet, and other means that is usually a push of the button away. Although this data has been good in countless ways, it has also created a world of Doubting Thomases who aren't quick to believe what they hear, much less see. Therefore, in order to make certain that Lean Manufacturing is taken seriously by the workforce, a plant's leadership has to demonstrate their intentions of making it a full and absolute success.

Doing so often requires a commitment to repeated communications regarding the merits of the process, along with the creation of incentives that serve to encourage participation. But it is important for a workforce to understand that although management is willing to listen to their doubts and concerns, the matter of Lean implementation is in no way negotiable. If properly accomplished, things will advance in a relatively smooth fashion. If not, the process is usually headed for trouble.

Appropriate discipline applies to insuring that the basic goals and objectives of Lean are fully carried out. In the real world, there are many things that can seriously distract, making it easy for management to overlook minor infractions. But it's very important that management doesn't allow this to happen — even once.

Employees are generally willing to work hard to see something accomplished if they believe steady progress is being made toward a clear set of objectives. But they're not as willing if the project is never ending by definition. People have to be able to see the light at the end of the tunnel, so to speak, and precisely where a specific level of accomplishment rests. Providing that direction is something that hasn't been done well in most Lean undertakings. Although I would be the first to oppose interrupting a Lean initiative for anything, there is a time and place for a temporary delay. However, a delay should not take place in the middle of efforts to achieve a specific level of Lean (e.g., Level One, Level Two). Working toward a clear set of objectives, which have an established start and completion point, makes the job of accommodating temporary delays much easier than would otherwise be the case.

Suppose a company has been in the process of actively implementing Lean in two of its three manufacturing facilities. The process was delayed at the third facility until it could finalize a new corporate initiative aimed at vendor certification. Near completion of the pilot, vendor certification proved to be such a success that many customers virtually insisted it be driven throughout the organization, in the fastest manner possible. Facing this dilemma, management felt it had little choice but to put its lean manufacturing effort on hold and actively pursue the new vendor certification program at all three plants. As a result, Lean never fully recovered. In fact, it came to be viewed by some as a complete failure.

With a slightly modified version of this story, the outcome would be entirely different. If Adams clearly understood where to put its efforts with Lean, in order to achieve a Level One Lean status, it could likely excuse the assigned personnel working on Lean from the vendor certification project and still proceed with both initiatives. In the worst case, it could delay vendor certification somewhat by letting customers know it was working on achieving the first stage of Lean Manufacturing, noting the sizable benefits it offered and proposing to start the vendor certification program at the other two factories as soon as the first stage was completed. This would make Lean less than a never-ending process in the eyes of the customer and open the door for a potential compromise.

GETTING THE MESSAGE OVER TO THE TROOPS IN A CLEAR AND UNDERSTANDABLE FASHION

For those embarking on a Lean undertaking, getting the proper message over to the workforce regarding the need for change is vital to success. In doing so, the message has to be clear with regard to intent. The message should also be communicated to employees by the highest level of management possible.

I would not be presumptuous enough to explain word for word what needs to be communicated. Still, I would point out some things which should be considered in the content of the message:

1. Emphasis should first be placed on making certain the workforce clearly understands the importance of the message and how crucial it is to the future of the operation. As an example:

   "I've asked you here today to speak about a matter that's crucial to the future of our operation. In doing so, I want to stress that we need to change the way we go about conducting business — not because of anything you have done wrong, but because the world of manufacturing is going through a very dynamic shift in thinking. We need to insure that we're among those who have come to recognize the need and act accordingly."

2. Effort should be made to describe the change. For example,

   "Like most manufacturers, we initially set our factory up to perform what's called 'batch' production. This was geared around each operation producing as quickly and efficiently as possible. It also allowed for doing so in a highly independent fashion, without a lot of coordination among all the activities in the factory. This approach worked fine as long as everyone in the industry was using the same technique.

   However, many manufacturers have begun to explore a new approach, one that better utilizes employees in the decision making process and greatly reduces both inventory and the space required to do the job. In doing so, they have come away with a much

more organized workplace, that makes the job easier to perform. Along with this, waste in the form of scrap, rework, and downtime have been greatly reduced, if not virtually eliminated.

The approach I'm speaking about is Lean Manufacturing. It's called that because it focuses on reducing wastes to the leanest amount possible. It also relies heavily on significantly reducing setup and changeover, along with improving layout and flow. And it's an approach we are going to apply to our plant.

3. It is also essential to note that management realizes the process isn't something that can be fully communicated in one short meeting. For example,

    "My purpose today is to give you an overview of our plans, not to spell out all the details regarding how we're going to go about making the change noted. However, I can promise you over the coming weeks and months, you will be hearing much more about the details, and you'll have an opportunity to raise questions and concerns you have. But let me assure you that this effort isn't something you have to fear. If we do things right, it's going to make your job better. In fact, one of the guiding principles of Lean Manufacturing says: 'If what we do doesn't make the job easier, then we're clearly doing something wrong!'"

4. Elaborate somewhat on the associated risks and benefits. For example:

    "If we choose to ignore the call and fail to respond adequately, our future could be at stake. I would be ter-

ribly remiss if I didn't point that out. But if we approach this effort with the proper frame of mind, we can potentially gain added business. We can not only insure job security, but enhance it, by cutting wastes and making our operation much more responsive to customer needs. In doing this, we can end up with a factory in which you can take considerable pride."

5. Summarize by stressing the depth of change and dedication required. For example,

   "Because this effort is going to take time, the manner in which you go about doing your job isn't going to change right away. But as things proceed, our methods of production will be largely refined. The way you perform your job will also change. I assure you the change will be for the better. I'm sincerely looking forward to our journey together. I ask that each and every one join me helping us make this next positive step toward our future."

The message clearly said a number of things, including:
1. Management recognizes the need to make a significant change.
2. There is adequate proof and good reason to do so.
3. Everyone has a stake in making the change a reality.
4. On-going communications will be vitally important.

There are also things the message doesn't say or promise:
1. The change will be easy and completely painless to

accomplish.

2. Management believes employees clearly understand the impact of the process on their future roles.

3. There won't be some unfortunate casualties along the way.

KEY REFLECTIONS

- Something better doesn't always mean something easy to implement; nor does it mean people will flock to be a part of it. There are struggles that go along with change and the bigger the change is, the more difficult the struggles become.

- In all likelihood, telling employees that Lean is a better way will not do the trick. They will generally put little stock in that, since "better" has been scrutinized and resisted throughout the ages

- Middle management can be one of the more serious stumbling blocks to a Lean undertaking, if not unquestionably the most significant barrier.

- What matters is whether a particular style of management will prove to have a positive or negative impact on the effort.

- One should avoid thinking that almost any conscientious employee, with reasonable communication skills and proper training and exposure to the process, would suffice to be the factory's Lean Coordinator.

- Keeping the operation on the right track and headed in the right direction is a primary task of the Lean Coordinator; these abilities usually require a special kind of individual.

- Without the strong commitment of the plant manager, the full incorporation of Lean is doomed from the start. He or she effectively represents the "conscience" of the process and, in the best case, employees perceive the initiative as a personal agenda of the plant manager,

- Employees are generally willing to work hard to see something accomplished if they can feel steady progress is being made toward a clear objective — but not if it's one that's never ending, by definition.

- Without a highly-focused maintenance effort, a Lean Initiative has little chance of success. Unfortunately, some maintenance managers are so steeply embedded in the status quo that they are extremely reluctant to accept a new and revised role for the maintenance function.

5

EXAMINING LEVEL ONE LEAN
IN CLOSER DETAIL

This chapter will examine Level I Lean in greater detail and build a better understanding about how to proceed in accomplishing it.

CHOOSING AND ALIGNING THE ENGINEERING STAFF

Proposing to focus on a number of qualified engineers to kick off a Lean initiative isn't the typical method that most operations have come to believe is the proper approach. But I would ask anyone who seriously feels that way to step back long enough to look at the merits of what I'm proposing. Just as important, I would ask them to consider the general lack of effective progress that many firms have experienced with a Lean undertaking.

It has taken two years of serious study and evaluation for me to draw any firm conclusions. However, I am convinced there is indeed a better way. In all fairness, one should be willing to look at the potential drawbacks of what they are proposing, and I'm more than willing to do that.

One drawback to focusing on the engineering staff is that it can serve to delay the training and involvement of the general workforce. I feel very strongly about training the workforce because it is absolutely crucial to long-term success. On the other hand, delaying their training until a plant's equipment is fully up to speed — as it pertains to the organization's ability to effectively support a Lean effort — could actually make employee training and involvement all the more successful. So as far as a potential drawback is concerned, I think it would be minimal.

Another potential drawback is that it narrows the starting scope of the process down to four specific tools and techniques (SMED, Poke-Yoke, Standard Work, and TPM). No work whatsoever would be done initially on other important elements of Lean such as Kanban, a formal pull system of production, or the use of one-piece flow techniques. The counterargument is that these elements can be much more actively applied once a plant's key equipment has reached a prescribed level of compliance.

Yet another potential drawback could be the difficulty associated with hiring, reassigning, and otherwise obtaining the engineering talent required to do the job noted. This could indeed pose a problem for those who do not have engineering staffs. However, this drawback isn't insurmountable, especially for firms that truly want to make the process more than a passing fancy.

As far as the qualifications of the engineers, the best fit

would be a team of industrial and manufacturing engineers who have a minimum of two-to-five years experience in manufacturing and who have demonstrated an ability to work with production operators and supervision in a Lean Manufacturing environment. The use of new college grads or those working in a co-op program should not be considered for this particular mission. This isn't to say they couldn't be useful, working under the direction of well-experienced engineers. However, it would not be wise to give them chief responsibility for the task.

To make this initiative all the more meaningful, special incentives should be considered. These could include a special bonus when the work was successfully completed. After all, you are essentially turning the framework for your entire Lean initiative over to a small group of talented people. Therefore, rewarding them accordingly would certainly seem appropriate.

ASSURING ADEQUATE MAINTENANCE SUPPORT

Experience has taught that adequate maintenance support typically ends up being one of the bigger stumbling blocks in a Lean undertaking. In the average batch-oriented factory, one or more of the following conditions usually exists:

1. Maintenance leadership is not fully aligned with world class maintenance practices.
2. Day-to-day upkeep of equipment and facilities is sometimes more than the existing maintenance staff

can effectively handle, much less take on the added responsibility of actively supporting a Lean effort.

3. The maintenance function, in general, has a "fix-it-when-its-broke" mentality.

The issue of having sufficient maintenance support for Lean is no small challenge. This is especially true for firms just embarking on a Lean Initiative. In many cases, outside maintenance will be required in order to rearrange equipment and facilities. This cost, of course, would be an extraordinary expense. But if all one desires is to make a cursory effort, in order to prove that the organization has at least ventured into the process, then maintenance support really isn't all that important.

On the other hand, for those who genuinely want to make Lean a lasting reality, the issue of appropriate maintenance support cannot be taken lightly. It can indeed become one of the more troublesome aspects of implementation. The need to clarify maintenance support is one of the chief reasons an 18–24 month Lean Implementation Plan should be prepared and approved by senior management very early in the process. Developing and implementing such a plan is the only way to realistically address the scope of the work to be done from a maintenance standpoint and the associated costs involved.

Assuming that a plan for maintenance support is developed and a decision is made to fund the work required, an individual in the maintenance ranks should be identified and assigned to the Lean Team, under a dotted line reporting arrangement to the Lean coordinator. Ideally this individual would be someone with a broad range of experience in maintenance. The job would be to work at coordinating maintenance activities as needed to meet the plan. In factories where this approach is taken, the assigned maintenance person becomes a vital asset in making a plant's Lean initiative more effective overall.

PERFORMING ADEQUATE TRAINING

One of the key elements of Lean is an on-going and effective training process, which needs to be structured with the goal that, at minimum, every employee will receive basic training in Lean. In approaching this task, four levels of training are proposed:

1. Lean Awareness
2. The Fast Track Event
3. The One-Week Special Event
4. The Two-Week Comprehensive Event

LEAN AWARENESS TRAINING

The first level provides basic training in Lean (see Figure 5.1). Every member of the workforce would be expected to complete Lean Awareness training. In addition, Lean Awareness should be built into new employee orientation. Some companies have taken it a step further by establishing an understanding that new employees will not become fully vested until they have successfully completed the training.

LEAN AWARENESS TRAINING

- *Introduction to World Class Manufacturing*
- *Understanding the Hidden Wastes*
- *Floor Exercise*
- *Report Out*

Notes:
1. *Intro to World Class Manufacturing is a one-hour overview of the Toyota Production System and its benefits verses typical batch manufacturing.*
2. *Floor Exercise is a team-oriented examination of the wastes, using Toyota's "Seven Deadly Wastes" as the guideline.*
3. *Report Out is a presentation by the team to management, covering findings.*

Figure 5.1

Lean Awareness training is a four-hour course. Participants are given a one-hour overview, followed by an hour aimed at "Understanding the Hidden Wastes." They are then asked to go to a select area in the factory and spend an hour, as teams, searching for and identifying the hidden wastes. The last hour of the course is spent listening to the teams report their findings and in answering any questions they may have. The training is completed with a brief review of what was covered and the importance of the participants lending their support to the process as the opportunity arises. Ideally, the training would open with a short video or presentation from the operations manager, who would stress the importance of the Lean effort and urge their participation.

FAST TRACK EVENT

The Fast Track Event is a three-day Lean training and implementation exercise (see Figure 5.2). It serves as the backbone for advancing the process throughout the factory. It is specifically aimed at both training participants and implementing Lean on the shop floor. The event is somewhat limited with regard to the depth of change undertaken because it is usually aimed at one particular manufacturing area, which belongs to a larger parent process. For example, the training event might be directed at a spot welding sub-assembly operation that is only one of a number of such processes making up the plant's Welding Department.

Figure 5.2

The course is structured to spend a day and a half on training. Some floor work is included during this period, identifying "the hidden wastes" and better understanding the man-

---

THREE-DAY FAST TRACK EVENT

Day One
- *Introduction to JIT*
- *Understanding the Hidden Wastes*
- *Push/Pull Exercise*
- *The Tools Of Lean*
- *In-depth Training in SMED, Poke-Yoke, TPM, Workplace Org., Other*

Day Two
- *Continuation of In-Depth Training*
- *Floor Work — Gathering Data, Information*
- *Establishment of Action Plan*

Day Three
The day is spent making changes on the shop floor, with a late afternoon report to management on results and work that needs to be done over the following two weeks to completely implement the change.

Figure 5.2

ufacturing process the team will be working on. With the help of the instructor, the team formulates a plan of action and the following day and a half is spent making the change on the shop floor. Before proceeding, however, the team's action plan is closely scrutinized. Based on the instructor's experience, a decision is made as to the actual degree of change the team will undertake. In most cases, this usually isn't a matter of having the team lower their expectations, but instead encouraging them to challenge themselves to an even higher level of accomplishment.

Once the work is completed on the shop floor, the established objectives and final results are presented to management by the entire team of participants. This presentation is followed by a factory tour to see first hand how the process was revised. This combination of presentation and tour serves to build teamwork and pride of accomplishment for those involved.

Once equipment has been brought up to minimum Lean standards, Fast Track Events should be on-going and continuous. They should be conducted by the Lean coordinator at least once every two weeks, throughout the course of the plant's 18–24 month Lean Implementation Plan, and they should be carried out without exception.

THE ONE-WEEK SPECIAL LEAN EVENT

The One-Week Special Lean Event is another training and implementation exercise for the factory, but as the name implies, it's very "special" in nature (see Figure 5.3). This event is most often conducted for something deemed espe-

Figure 5.3 One Week Comprehensive Event

Classroom and Floor	Day 1	Day 2	Day 3	Day 4	Day 5
	*Opening	*Intro to SMED	*Visual controls	FLOOR WORK MAKING CHANGE	FLOOR WORK MAKING CHANGE
	*Intro. To JIT	*Intro to Poke-Yoke	*TPM		
	*Intro. to Waste-Free Mfg.	*Understanding the hidden wastes	FLOOR EXERCISE (Visual controls)		Presentation to Mgt. and tour to show changes made
	*Presentation on the situation/problem to be addressed	FLOOR EXERCISE (hidden wastes)	Establish action plan		

cially important for the factory, such as the addition of a new assembly line to support added business or a new piece of sophisticated equipment, in order to see that the production process is set up in accordance with good Lean practices. The event can also be used to address a serious problem that has developed on a major manufacturing process in the factory, pertaining to poor quality, throughput and the like.

This event usually carries a high level of exposure and should consist principally of participants who have been through at least one "Fast Track" event. The engineers and shop floor mangers who are involved on a day-to-day basis with the process in question should also be included. The purpose is to gather the people who are most familiar with the equipment and process, along with others who have applied experience in Lean. Combining first hand knowledge with fresh thoughts and ideas will typically result in a solid plan of action and better results overall.

The first day of the event is spent on a general review of Lean, something most of the participants may have already heard, but which is needed in order to get them thinking in terms of how to use Lean concepts, principles, and techniques in addressing the issue. This review of Lean is followed by a presentation conducted by those participants most familiar with the equipment. They will give an overview of the situation the team will be facing and eventually provide a solution.

The first half of the second day is spent on advanced training in the tools that will be used to provide a Lean influence on the final solution. These tools may include SMED,

Poke-Yoke, TPM, or some combination thereof, as required. During the latter part of the second day, the team is sent to shop floor where they stay the remainder of the event, until tasks are completed as planned. The only exception is returning to the training area at the end of each work day to review progress and to discuss issues and strategy, as required. This overall process very often calls for the team to work over the weekend, in order to see everything is fully completed. As before, at the end of the event, management reviews the results and tours the plant to view the progress/solution made.

As plant manager, I was there to help kick things off and returned for the final management review. But I also took the time to attend the afternoon reporting sessions and went even further by asking my direct reports to accompany me and participate in the discussion, as needed. These steps indicated to everyone involved that the work being carried out was deemed highly important to the plant's senior management staff.

THE TWO WEEK COMPREHENSIVE LEAN EVENT

This event usually happens only once over the course of a plant's journey into Lean (see Figure 5.4). It serves a two-fold purpose. The first purpose is to train and involve a highly select group of individuals. This group consists principally of key managers and other operational personnel, some of who may have no direct tie to the manufacturing floor, but who are indirectly or otherwise impacted by the process. Also

Figure 5.4 Two Week Comprehensive Event

WEEK ONE	Day 1	Day 2	Day 3	Day 4	Day 5
CLASSROOM AND FLOOR	*Opening	*Intro to SMED	*Std. Work	*Revisting: - JIT - SMED - Poke-Yoke	*One Piece Flow
	*Intro. to JIT	*Intro to Poke-Yoke	*Visual controls	*Ridding Your Paradigms	*Exercise: (Pull Vs. Batch)
	*Intro to Waste-Free Mfg.	*Understanding the Hidden Wastes	*TPM	FLOOR EXERCISE (Gather Data)	*Kanban
	*Intro. to the Hidden Wastes	FLOOR EXERCISE (Hidden Wastes)	FLOOR EXERCISE (Std. Work)		*Forming Plan of Action
					*Present Plan to Management and Obtain Buyoff

WEEK TWO	DAY 1	DAY 2	DAY 3	DAY 4	DAY 5
MAKING CHANGE ON THE SHOP FLOOR	FLOOR WORK	FLOOR WORK	FLOOR WORK	FLOOR WORK	FLOOR WORK
	*REPORT PROGRESS	*REPORT PROGRESS	*REPORT PROGRESS	*REPORT PROGRESS	*MGT. PRESENTATION

included, if possible, are two-to-three participants from other factories within the company and at least one key supplier representative. It has been my experience that participants from outside the factory bring fresh thoughts and ideas that might otherwise be overlooked. This is because they haven't been influenced by the way business has been conducted at the plant in question and come with an open mind.

Key managers should include the plant manager, along with the engineering managers and others who play a role in day-to-day production. Supporting staff functions should include personnel who do not normally report to production, but are functionally a member of the factory team. These include such functions as Quality Assurance, Accounting, Human Relations. Purchasing, Materials, and Production Control. Other key operational personnel would typically come from functions such as Sales and Marketing, Product or Design Engineering, and Systems.

The second purpose of the event is to make significant change on the shop floor. In fact, the entire second week is dedicated to the effort, except for a short report on program at the end of each day. The event is geared at getting the message across to those who sit in some very influential positions throughout the company. In total, the number of participants involved should be kept to a maximum of thirty in order to keep training as individual as possible. With anything more than thirty, the event tends to become more of a lecturing session than needed interplay. On the other hand, anything less than twenty participants for this particular event would likely be too few to do the considerable floor work required.

Getting a solid commitment from a key manager to spend two weeks in such an event is often next to impossible, although I have seen situations where it was insisted at the highest level. Even though getting the department manager to attend isn't likely in most cases, every effort should be made to have the highest ranking individual possible from each of the functions noted.

It definitely requires a very strong personal commitment for someone to take two weeks out of a busy schedule and dedicate it exclusively to a special training and implementation session — in this case, a full education into the inner workings of Lean Manufacturing. But at least one such event should be conducted at any company seriously pursuing Lean.

As mentioned, this event is extremely comprehensive in nature and involves making substantial change on the shop floor. Therefore, the maintenance department has to be aligned to support the effort. Making substantial change may indeed require going outside for help, in order to get the job done. The cost for such an event can be notable, but the rewards can be gigantic in terms of getting the message across and building needed support. The company as a whole, and certainly those in functions that are not directly tied to production, must come to recognize the considerable advantages of Lean and the gigantic amount of work required at a manufacturing level in order to put needed change in place and maintain it on a long-term basis.

THE ORDER IN WHICH WORK SHOULD BE PER-FORMED ON A PLANT'S KEY EQUIPMENT AND WHY

Earlier I noted the basic work that needs to be done in order to bring key equipment up to compliance and achieve a Level One Lean status. But there is a need to address this in greater depth at this point and provide reasons why it's important to proceed in an orderly and disciplined manner. The tasks, in the sequence of order as this applies to the work performed by both the industrial and manufacturing engineers, are as follows:

1. The application of SMED (ME) and Standard work (IE)
2. The application of Poke-Yoke (ME) and Workplace Organization (IE)
3. The application of TPM (joint ME/IE effort)

The I.E. and M.E. work together as required and both assist in addressing a particular issue. But it is very important to maintain a reasonable level of discipline to the order of work involved. Most important, do not proceed to the next individual task until the work of both engineers is fully completed for the step noted.

The reason we want to see SMED and Standard Work completed simultaneously is SMED establishes the new method of changeover and Standard Work effectively measures and records both the old and new method, with an outcome being adequate work instructions. This, of course, can't be done effectively until SMED is fully completed or vice versa. Therefore, if one of the engineers hasn't finished work on a particular piece of equipment, the other engineer would lend assistance, until the entire first step was completed.

Another reason Step One should be carried out in its entirety before advancing to Step Two is Poke-Yoke (or Mistake Proofing) cannot be effectively applied until the new method utilized to set up the equipment is clearly established. The same holds true, in most respects, for the full completion of Workplace Organization and TPM — although admittedly, some aspects of each will take place throughout the entire process.

On advancing to the second step, the I.E. should pursue the completion of Workplace Organization while the M.E. is working on Poke-Yoke. Once completed, both then proceed to step three and involve the operators and the maintenance function in establishing a TPM for the equipment.

Depending on the complexity of the equipment, two-to-three weeks should suffice to complete the work on the first key piece of equipment. But as the engineers gain both experience and momentum, the time will gradually improve.

Assume a plant has 300,000 square feet of manufacturing space and 100 individual pieces of equipment, with two teams of engineers working on the project. Seventy of those individual pieces are classified as Key Equipment and of these, 20 are relatively complex in nature. Because the 20 noted will in all probability be the first ones tackled, it is safe to assume it will take an average of 10 working days per machine, plus 5 days per machine for each of the following fifty machines — although they might not require that long on average. Using this as a rough calculation indicates a total time of 225 days required to fully complete the task.

Most operations plan a 240-day-a-year schedule (48 weeks X 5 days a week). Therefore, with a team of four well-qualified and energetic engineers, the project could be completed in just under a year. The addition of one extra team (making it six engineers in total) would bring the time span down to approximately 6 months or less. However, it all boils down to how quickly an organization wants to have the project completed and how much of a dedicated engineering staff they can afford to apply.

The next question is if it is possible to have someone other than a trained engineer working on certain elements of the project. The answer is yes, although care should be taken. Others can indeed be assigned to perform some of the work if they clearly display the ability. On the other hand, caution has to be taken when looking at this option, especially when it comes to the matter of SMED and Poke-Yoke.

An operation pursing a Lean One Lean status has to carefully analyze the situation and try to determine the necessary resources required to do the job in a reasonable time-frame. A year for this effort isn't excessive. On the bright side, the rewards of successfully performing the work will almost immediately began to show in the plant's ability to be more flexible in meeting customer demands. Indirect labor costs such as setup and changeover will also begin to show some remarkable improvement. The quality of the parts and components should be of much better quality, thereby reducing scrap, downtime, and rework.

THE APPLICATION OF SMED

Taiichi Ohno's book *Toyota Production System* should be required reading. There is nothing else related to Lean that is as simple in structure, yet as powerful in content. In discussing setup, Ohno points out that it was looked upon in Toyota as *"an element that reduces efficiency and increases cost."* He goes on to say, *"There seemed no reason for workers to want to change dies cheerfully. Teaching workers to reduce lot sizes and setup times took repeated on-the-job-training."*

Ohno's remarks are interesting, especially because something commonly heard is that the Japanese worker has a totally different attitude about things. Yet Ohno indicates they had no less of a problem changing the attitude of employees in Japan than we have in the United States. What Toyota did have, however, was some very strong convictions about the need for change and a strong determination to see it through.

We can't forget the fact that Toyota, like most industry in Japan after World War II, initially set up its factories to produce in a batch format, much like the United States. What drove Toyota to change was an inability to compete in a world market, due to a gross lack of natural resources. They were determined to find a better way and were very smart in how they went about it.

Most notably, they took Henry Ford's philosophy and energetically applied the basic concepts he preached. They had to teach the workforce to change, just like anyone heading into a Lean initiative today. Ohno coupled the vital importance of reducing setup to Toyota's "Mixed Model Production and Leveling" concept that eventually came to set the U.S. automotive industry on its heels.

Ohno used the term "quick changeover." I personally prefer his definition better than SMED (Single Minute Exchange of Dies). Setup and changeover encompass much more than changing dies. Even more important, the concept of quick changeover isn't about meeting some specified span of time, but to work continually at bringing the setup time down to the point that it isn't a factor in the decision making process. The reason setup reduction was called *Insignificant Changeover* in *Fast Track to Waste-Free Manufacturing* was to place awareness on this very point.

The matter of definition is also why many firms quickly give up on SMED or never start it in the first place. It is one thing to have an objective of working actively to reduce setup and quite another to get every setup down to nine minutes or less. The time itself isn't as important as the influence setup has on making good business decisions (increasing a line rate to take on added business, etc.). If a particular setup gets in the way, then it's time to reduce it further. In some cases, bringing a setup down from two hours to thirty minutes can be just as effective for the business as nine minutes or less.

An important aspect of tackling setup times is utilizing good work measurement (Standard Work). It was Shingo who defined effectiveness at work measurement as "observing the smallest tasks." Every element of a setup needs to be measured and recorded professionally. These observations can serve as the basis for continuous improvement —it is much easier to know where you are going when you fully understand where you are. Work measurement provides the means of doing this. If performed in the proper manner, Standard Work points out every step involved in what is described as "the latest method being utilized" and points to potential areas for further improvement. Please note that the "very best" method isn't mentioned; the very best method can only come about with repeated efforts, if ever totally achieved.

SMED and Standard Work go hand in hand helping to raise a plant's key production equipment up to the level required to adequately support Lean Manufacturing. The intent of this book isn't to provide the specifics regarding how to do either SMED or Standard Work. There are already numerous resources on the market relating precisely how to apply the tools. Furthermore, I have assumed the reader has a general working knowledge of their application. More important is building an understanding about the approach to take and what tasks to tackle first.

Once SMED and Standard Work have been fully applied to all key production equipment, the stage is effectively set for advancing some of the more important aspects of Lean, such as Mistake Proofing (Poke-Yoke) and TPM (Total Productive Maintenance). In doing so, the factory reaps immediate benefits in terms of lower indirect labor costs and far less downtime, scrap, and rework.

THE APPLICATION OF WORKPLACE ORGANIZATION AND POKE-YOKE

In *Fast Track,* Workplace Organization is mentioned as the foundation for all continuous improvement. I still very strongly feel that way. However, in the scope of this particular approach to implementation, I would place it second on the list of things to do. I say this because Workplace Organization is much easier to accomplish once a viable method of setup has been established, adequately measured, and fully implemented. The workplace can then be organized to fit the prescribed method, thus reducing the need for trail and error.

Mistake Proofing — or the use of techniques inherent to administering Poke-Yoke — is also addressed at this point. There are some who would say SMED and Poke-Yoke could be done in conjunction. But this disregards the importance of placing structure on a very specific task until it is fully completed. Combining the incorporation of quick changeover with mistake proofing leaves far too much room for placing attention on somehow completing both tasks and subsequently doing a less than adequate job of either.

Applying discipline to the task of equipment engineering involves following a clearly defined path of accomplishment and staying with the job until it's fully completed. This can sometimes be problematic, especially with engineers who have a good background in manufaction and see an opportunity to be helpful to others (such as stopping by to help get a troublesome piece of equipment running again, assisting with a feeder jam and the like). A fine line has to be walked between discouraging good intentions and insuring that the stated mission is fully accomplished. It will usually fall to the Lean Coordinator to see that things remain on track and do it in a manner that doesn't come across as overly critical. A tough job, to say the least.

As mentioned, we will not specifically get into how to apply Poke-Yoke. However, the work should be performed by a qualified engineer who has a good background in equipment installation, specifications, and the like.

THE APPLICATION OF TPM

Once a plant's key production equipment has had SMED, Standard Work, and Poke-Yoke fully applied, it is time to focus attention on preventive maintenance and upkeep. Ohno speaks about "Taking Care of Old Equipment" and could just as easily have said taking care of *any* equipment. He notes,

"The language of business economics talks to depreciation, residual value or book value — artificial terms used for accounting, tax purposes, and convenience. Unfortunately, people seem to have forgotten such terms have no relevance to the actual value of the machine. If a piece of equipment purchased in the 1920's is kept up and can guarantee, at present, an operable rate close to 100% and if it can bear the production burden placed on it, the machine's value has not declined a bit."

The purpose of TPM is to go as far as possible in seeing that every piece of equipment considered "operable" can bear the production burden placed on it. Some manufacturing firms would classify a machine operable if it somehow managed to limp along, far short of expectations, in terms of both output and quality — in other words, if it was capable of producing something, even through it fell short of purchase specifications. Toyota, on the other hand, would find this totally unacceptable. It would shut the equipment down until it was brought up to good measurable standards.

The difference in mind-set is essentially the difference between a world-class manufacturing operation and those that simply tag along. I have had people tell me that it's nonsense to consider shutting down equipment when the problem can be addressed at a time when production isn't affected. What they are blinded to is the fact that it *does* affect production, and in a very serious way. Once a firm implies it's okay to operate a machine that isn't functioning to its established specifications, it is effectively telling the workforce that it's acceptable to *deviate.* In turn, once a workforce sees it's acceptable to deviate, a *precedent* has been set from which it is often difficult to recover.

In *TPM for America — What It Is and Why You Need It*, authors Herbert R. & Norman L. Steinbacher write, *"American Manufacturers regrettably have not supported maintenance — the very foundation of industry — while demanding ever greater performance from it. Without TPM, neither just-in-time nor quality management exist — at best they are dreams."*

In their book, the Steinbachers note five levels of advanced maintenance:

1. Prevention Maintenance
2. Predictive Maintenance
3. Corrective Maintenance
4. Preventive Maintenance
5. Autonomous Maintenance

The first four levels have to do with keeping equipment running as needed, by designing and selecting the appropriate equipment, keeping it properly maintained, and understanding the inner workings well enough to predict and prevent failures. The last level, Autonomous Maintenance, has to do with involving production employees in the maintenance process. I would highly suggest the Steinbachers' book to anyone seriously considering Lean Manufacturing.

CONDUCTING THE PLANT'S FIRST HIGH-LEVEL TRAINING SESSION

The first high-level training session, aimed at making a significant change to a given area of the factory, comes next. Although this session shouldn't be considered the single most important training given, it is unquestionably the single most important training *event*. It sets the stage for both the type and the depth of change that workers are going to see carried out throughout the factory. Therefore, it needs to be approached as a very special occasion.

George David, the President and CEO of United Technologies, who was recognized in 2005 as one of the best CEOs in the nation, was a firm believer in Lean Manufacturing and adamant about the need for training. He strongly believed it had to go further than simply involving plant personnel. He insisted that the highest level of executive management within the corporation receive hands-on experience. The power and influence of that approach was awesome, to say the least; out of it came some remarkable accomplishments.

Although this kind of senior management commitment is extremely difficult to find, it doesn't negate the fact that the first major training event, designed to show the power of the process, should be as extensive as possible in scope and include key individuals outside the manufacturing arena.

In Chapter Four I spoke about the Two-Week Comprehensive Lean Event and suggested a list of participants. I should stress that all participants are expected to perform in a hands-on manner. During the event, their positions and titles should essentially be forgotten — every participant should be considered of equal status. The objective should be to learn and work together toward making appropriate change.

Granting all participants equal status can sometimes be a difficult thing for the higher-level participants to accept and energetically work at accomplishing. However, given the proper understanding and encouragement from top management, it can be done. The reason I know is because I have personally seen it happen.

Selecting the pilot area is an important exercise in itself. It should be a location in the plant that operates as closely as possible to actual customer requirements. Final assembly processes are typically ideal. Regardless, the area should be one that is creating some of the biggest problems for the factory in terms of cost, quality, and delivery. Due to the nature of the event, it marks one of the rare times a factory will have all the clout and influence required in one team. Therefore, it is important to take advantage of this opportunity and to make the kind of change that is both significant in scope and lasting in effect.

A WORD OF CAUTION ABOUT INTENT

Allow me to mention something pertaining to intent. As the steps spelled out in achieving Level One Lean are made, it should be obvious that this effort is very serious business. It is a tremendous commitment for an operation to make, and if the intention of management is anything less than one that aggressively applies Lean across the factory as a whole, my advice is to think twice before starting.

I have seen operations that ventured into Lean with something less in mind. When this happens, they are effectively misleading themselves, the workforce, and ultimately their customers. To make Lean Manufacturing work, an operation has to be inspired to change completely and thoroughly. Otherwise, it's an exercise in futility and can actually do more harm than good.

THE 18–24 MONTH LEAN IMPLEMENTATION PLAN

Once the initial Two-Week Comprehensive Lean Event has been successfully accomplished, work should be done to put together an 18–24 Month Lean Implementation Plan, encompassing the entire factory. Some might see this as something that needs be done before this event, but there is good reason why it should be afterwards.

The Two-Week Comprehensive Lean Event will tell a factory much about what it needs to do in order to make Lean Manufacturing a way of life. It will point out, among other things, how good a factory is going to be with needed maintenance support, how well its people buy into the process, how various shortcomings that were experienced will come to bear on the process, and which back-up processes in the factory need immediate attention.

Three items should be clearly addressed in the plan:

1. The precise order each section of the factory will be covered and why
2. The number and type of training events scheduled
3. The estimated cost of full compliance to the plan

The plan requires some extensive work on the part of those responsible for its preparation. The effort, of course, should be headed up by the plant's Lean Coordinator. But it should also include the input and assistance of a number of key factory managers. It is best for the group to meet off-site, away from day-to-day distractions, because initial planning will normally require a couple of days of intense debate and consideration.

The plan should then be reviewed with senior management and formally approved. After the initial management review, certain aspects of the plan will likely need to be revised and resubmitted for approval. But once accomplished, the plan should be clearly communicated to the workforce in the most effective manner possible. This communication could include actions such as posting the highlights of the plan in every production area and consistently making reference to progress as the opportunity arises — along with utilizing the company's monthly newsletter or some other form of printed document to spread the word.

ESTABLISHING "OWNER-OPERATORS"

The last piece associated with fully implementing Level One Lean involves the establishment of "Owner-Operators" on at least 50% of the key production equipment in a factory. Anything less than 50% would not serve as an effective commitment. Anything more would usually take far too long before proceeding to the next level of accomplishment.

"Owner-Operators" are machine attendants who receive training and express the ability to perform general maintenance and upkeep of their equipment, as well as run production requirements. They should be compensated accordingly, which brings us to the matter of pay-for-skills-earned.

Pay-for-skills-earned is an element important to the overall success of Lean. It is especially important when it comes to Owner-Operators, which should be the highest paying job on the shop floor (or at least among the very highest). A plant doesn't want good, skilled Owner-Operators tempted to accept a higher paying classification at the first opportunity — especially because a great amount of special training would have been invested in them.

Again referring to *TPM For America,* the following is noted about operator involvement:

"Autonomous maintenance does away with lackadaisical attitudes once machine operators get past the 'it's not my job' syndrome. Machine operators are given responsibility for the equipment and for production quality; they are the ones who check production rather than someone at the end of the process; and they now have a greater level of accountability

for their work. Machine operators also have a greater level of authority and freedom to make decisions…. Moreover, machine operators can tell more about the condition of a piece of equipment than anyone else in the company. The regular driver of a car is familiar with its noises, vibrations and temperature. Any change would be recognized earlier by that person than by an occasional fill-in or casual observer."

What is clear, I hope, is that maintenance of equipment isn't complete until it goes much further than fixing a machine when it breaks. Given the knowledge and techniques available today, the fix-it mentality really isn't maintenance at all, but an avoidable mistake that should be viewed as such. Involving the operators is key. We have to face the fact that their involvement can't be achieved without effective training and appropriate remuneration. The money spent to accomplish this will more than pay for itself.

If any special care has to be taken, it is with regard to what qualifies a production employee to become an Owner-Operator. The answer isn't always simply the person who commonly runs a given piece of equipment. To qualify, a person should be tested as to ability to read prints, equipment specification sheets, etc., along with passing some form of mechanical aptitude test. Actually, these are tests that a good machine operator should be capable of doing in the first place.

I'm certainly not one for added labor classifications, but adding responsibilities of the magnitude we're speaking can sometimes warrant this. If a company takes the position that all machine operators have to become competent in the prescribed maintenance of their equipment, as well as the quality of the parts produced, it faces the chance of having to provide added compensation for all operators. That's separate from any flack it could receive because of someone who was totally incapable of performing the work prescribed. This is an excellent reason for the creation of Owner-Operators. Few labor unions could or would argue that meeting the noted requirements would not require special skill and aptitude. However, when you try to force new standards on long-standing classifications, you began to find employees who simply can't comply. At that point, you may start to get strong resistance from the union — and perhaps rightfully so.

KEY REFLECTIONS

- One of the key elements of Lean is an on-going and effective training process, which needs to be structured with the thought in mind that each and every employee will at a minimum receive basic training in Lean.

- Experience has taught that adequate maintenance support typically ends up being one of the bigger stumbling blocks; which generally boils down to three specific issues: maintenance leadership, maintenance workload, and a "fix it when it breaks" mentality.

- The Fast Track Event is a 3-day Lean training and implementation exercise. It serves as the backbone for advancing the process throughout the factory.

- The One-Week Special Lean Event is another training and implementation exercise for the factory. As the name implies, it is "special" in nature.

- The Two-Week Comprehensive Event usually happens only once over the course of a plant's journey into Lean. The purpose is to train and involve a highly-select group of individuals, principally consisting of key managers of supporting functions, some of whom have no direct tie to the manufacturing floor.

- Once the initial Comprehensive Lean Event has been successfully accomplished, work should be done to put an 18/24 Month Lean Implementation Plan together, encompassing the entire factory.

- When a plant's key production equipment has had

SMED, Standard Work, WPO, and Poke-Yoke fully applied, it is then time to focus attention on preventive maintenance and upkeep of the equipment.

- The last piece associated with fully implementing Level One Lean involves establishment of "Owner-Operators" on at least 50% of the key production equipment in a factory.

6

THE SEARCH FOR GREATNESS

THE NEED FOR EFFECTIVE LEADERSHIP

For the period starting after World War II and running through the mid-1980s, the United States reigned as the recognized global leader in manufacturing. Today we hold no special prominence in the field. The contributing factors are complex and involve both internal and external issues. However, one that can be attributed entirely to manufacturing was the ongoing resistance to accepting much more efficient ways of doing business, which would have led to improved quality and greater production flexibility.

I believe this resistance has had a much greater impact on the overall decline in U.S. manufacturing than it's been given credit for. Much blame has been placed on free trade agreements or lower labor costs in emerging economies around the world. These have unquestionably played a role. But, for the most part, the United States ignored the call for change until it was far too late. The best evidence of that is the fact that batch manufacturing still reigns on a relatively large scale in U.S. industry.

As more and more emphasis has been given to Lean Manufacturing, the awareness level has risen and we have chipped away at the problem. But there isn't a universally accepted method of approaching the task. To a large degree, we find ourselves floundering; unable to free ourselves of a 1940 model in a 21st-century world.

For those who ask what makes Lean Manufacturing any more important than any other initiative a manufacturing firm might undertake, the answer is simple — Survival! Yet the manufacturing paradigm that poses the question in the first place is extremely complex.

The system of production that largely influenced U.S. dominance and later proved to be its downfall was a model fashioned after the Henry Ford approach. In this model, everything was aimed at being highly productive in terms of output. Plant efficiency was gauged on how busy the factory stayed, regardless if what was being produced was needed to fulfill immediate customer requirements. Large volumes of inventory were viewed as an asset. Inspection and rework reigned as the gatekeeper for quality and factory measurements were geared to keep the production machine plugging away. Sound familiar?

Even for plants who expound to a Lean philosophy, the remnants of the 1940s model can still be detected. Certainly some of the more obvious wastes inherent to the process have been addressed. But finding factories that have learned to utilize Lean Manufacturing to its fullest and have entirely driven a push system of production out of their factory can sometimes be a challenging search.

If the 1940s model is examined closely enough; it will actually be found that the model isn't consistent with Henry Ford's approach. In fact, Ford would probably have been shocked with the wastes generated by most modern day manufacturing techniques. Henry believed in wasting nothing. For example, the wooden pallets on which raw materials were shipped to his factory were transformed into running boards for the automobiles he produced. In addition, Ford believed it was every operator's duty to inspect the work of the preceding operator and make certain the quality of the work was right. If there was problem, the line was stopped and necessary corrections were made.

Somewhere in the process of making batch manufacturing what it is today, someone of influence decided the line would run faster if inspection was taken away from the operators and given to a quality control function. It was also decided operators would be free to produce more if tasks such as lubricating equipment and sharpening cutting tools were given to a professional maintenance department. In addition, someone took Henry's idea of conveying a heavy product like an automobile from one station to the next and decided it was appropriate for assembly work of any kind. However, the objective wasn't as much weight related as it was the pacing the output of the operators.

Over time, emphasis slowly shifted to a "more is always better" mentality. As a result, special production support functions were established and overhead costs steadily increased, along with scrap, rework, and obsolescence. Companies began to view the hourly production worker as an easily expendable resource. In turn, this view created a framework for what was generally expected of them: Keep your nose clean, show up for work on time, check your brains at the door, and do your job. Sound familiar?

Because of an ever-increasing focus on high-volume production, an atmosphere was created which divorced shop floor employees from almost any form of decision making. I'm referring to those who work with production processes on a day-in and day-out basis and typically know the equipment they work with better than anyone else. But even today, the mindset of many production managers is that operators need to stay busy producing parts and assembling units — and decisions should be left to management.

To make a long story short, Toyota read Henry Ford's book, listened to Juran and others, and came to the United States to observe our manufacturing processes first hand. They found that what Ford aspired to and what manufacturing in the United States had become were two different things. With that knowledge in hand, the springboard for the Toyota Production System was born.

Nothing could be more critical to long-term success than overcoming the shortfalls mentioned. Batch production is a thing of the past. The sooner we collectively work to eliminate the practices that are making us less than fully competitive, the better off U.S. manufacturing will be. In order to fully achieve this, however, we have to rethink the whole matter of lean implementation.

A very high percentage of Lean undertakings in the United States have been pursued for all the wrong reasons:

1. Lean was viewed as the fashionable thing to do. Therefore, it was given a level of commitment that was no more than any other assigned strategy, and often quite less.
2. Someone in the middle management ranks raised the idea and senior management agreed to go along. Because it became an objective driven from the middle up, instead of the top down, the support and dedication applied to the process was, at best, less than enthusiastic.
3. An outside influence — such as a lending firm — stipulated Lean practices had to be applied. In order to adhere, it was agreed to make something happen. But for all intents and purposes, it was a forced initiative to which management felt no real obligation, outside of doing enough to appease auditors.

At the time of this writing, the United States is currently going through the most critical financial crisis it has faced since the Great Depression. Much of the problem is the continuing decline in America's manufacturing base. For Lean to be totally effective, we are going to need more leaders with the mindset of Toyota's Taiichi Ohno.

In the early 1950s Toyota was on the verge of bankruptcy, finding itself unable to afford major investments in new equipment or massive inventories. Ohno developed many improvements while working as Toyota's assembly manager. Later, in collaboration with Shigeo Shingo, he refined his efforts into an integrated manufacturing strategy, which today is known as the Toyota Production System.

Ohno's approach to inserting the Toyota Production System was largely a "my way or the highway" style of management. Critics often saw him as domineering, arbitrary, and difficult to work with. In his defense, Norman Bodek, President of PCS, Inc., and the man who was chiefly responsible for translating Ohno's book in English, noted in an interview with Strategos, "An employee has never been laid off, a supplier has never gone bankrupt, in fact each supplier is a leader in their field, and Toyota is able to open new plants in America while we do so in China."

In discussing Ohno's management style, Bodek said: "One day Ohno walked into one of the warehouses at Toyota Gosei and said to the staff of managers around him, 'Get rid of this warehouse and in one year I will come back and look! I want to see this warehouse made into a machine shop and I want to see everyone trained as machinists.' Sure enough, one year later that building became a machine shop and everyone had been retrained." But Bodek went on to clarify, "Ohno knew the economic benefits to Lean, knew it wasn't easy to bring change, and was forceful in bringing it forward."

Bodek also mention that he was later invited to look at a newly-purchased automated delivery system in a factory and was asked by one of the managers standing by what Taiichi Ohno would have say about it. Bodek replied: "Ohno would say get rid of it."

Bringing effective change at a manufacturing level is best achieved when leadership at the top of the organization soundly supports the change. Even better is when they are personally driven to make it happen. On the other hand, when it comes to Lean, this support cannot be relied on in most cases. All the more reason for a clear and structured path to implementation.

THE BEGINNER'S GUIDE TO EFFECTIVELY INSERTING LEAN

It's safe to say that anyone thinking about Lean would be looking for a process that would pay the strongest return

Figure 6.1

<u>THE NINE INITIAL STEPS TO SUCCESS:</u>

STEP 1. Select and establish a fully qualified Lean Manufacturing Coordinator (ref. chapter Five: "Picking Appropriate Leadership").

STEP 2. Reorganize and staff as needed to support the effort (ref. Chapter Five: "Choosing and Aligning the Engineering Staff).

STEP 3. Devise an 18-Month Implementation Plan and obtain top management buyoff (ref. this chapter).

STEP 4. Provide assigned engineers with an advanced level of training in SMED, Poke-Yoke, Standard Work, and Workplace Organization (ref. Chapter Two, Figure 2.1).

STEP 5. Communicate and provide a Lean Overview for all key managers (ref. Chapter Two, Figure 2.3 and Chapter Four "Getting the Message Over to the Troops").

STEP 6. Engineer equipment to support a Lean effort and while this is in process, train key hourly and salaried employees in the basics of Lean (ref. Chapter Two , Figure 2.2 and sections for Smed, Poke-Yoke, and TPM outlined in Chapter Five).

STEP 7. Once equipment has been fully engineered, select the pilot area and conduct the plant's first high level training and implementation event (ref. Chapter Five: "The One-Week Special Event" and "The Two-Week Comprehensive Event").

STEP 8. Start the process of on-going shop floor changes, involving teams of both hourly and salaried personnel (ref. Chapter Five: "The Fast Track Event).

STEP 9. Proceed with the items outlined in Chapter Three to achieve a full Level One Lean status.

on investment and effectively set the course for the future. Key to this approach is understanding the order in which work should be done, leading up to fully achieving a Level I Lean Status, and focusing efforts beyond.

The Beginner's Guide is specifically designed for those just entering into Lean. It's also a good reference for those already into the process who see the need to change strategy. The nine steps covered have been addressed, in varying degree, throughout this text (see Figure 6.1).

Steps 1 and 2

The first two steps are spelled out in Chapter Five, under *Picking Appropriate Leadership* and *Choosing and Aligning the Engineering Staff.*

Step 3

One of the more significant steps in achieving Level One Lean involves the development of an 18–24 Month Implementation Plan and seeing it through to completion. I personally prefer a plan that looks out over 18 months, but going up to 24 months is acceptable.

Preparing this document and getting a buy-off from senior management is vital. Far too often, this type of planning is overlooked in starting a Lean initiative. The actions outlined in Chapter Five, (*Establishing Key Equipment* and implementing SMED, Standard Work, Poke-Yoke, and TPM) should make the job of constructing the plan easier than starting from scratch and trying to figure out what should be done.

Once every six months, the plan should have a high-level management review; changes should be made, where applicable. A word of caution: strive to make any change to the plan one that serves to enhance, rather than distract from, initial goals and objectives. Making a change which lowers initial expectations should never be taken lightly. Every effort should be made to find a way to meet the original objective if at all possible. Remember you are on a journey. Every delay getting there opens a path of opportunity for the competition.

Step 4

The information needed to take Step 4 is spelled out in Chapter Two and includes a window diagram (Figure 2.1) that scopes the training on a day-to-day basis.

Step 5

Communications are covered in depth in Chapter Four under *Getting the Message over to the Troops*.

Step 6

At the same time that the assigned engineers are working on the plant's equipment, and in order to support Lean more effectively, every effort should be made to give all employees *Lean Awareness* training. The outline for this training is described in Chapter Two, Figure 2.3. The specifics regarding the assigned work of the engineers is covering in Chapter Five, under *SMED*, *Poke-Yoke*, and *TPM*.

Step 7

Once the equipment has been brought up to minimum standards, the plant is ready to select a pilot area and conduct a *One-Week Special Lean Event* or a *Two-Week Comprehensive Lean Event*. The type of event depends on the depth of work needed to make the pilot a success. How to go about performing both events is spelled out in Chapter Five.

Step 8

Getting the workforce started in making change is critical to the process. This effort should involve teams of both hourly and salaried personnel, in a setting designed to couple needed training with the implementation of Lean concepts on the shop floor. The procedure required is spelled out in Chapter Five, under *The Fast Track Event*.

Step 9

Everything necessary to meet a *Level One Lean Status* is covered in detail in Chapter Three.

SPREADING LEAN INTO SUPPORTING BUSINESS FUNCTIONS

This next step essentially involves performing an *Office Kaizen* in key supporting areas of the business, including Purchasing, Accounting, Production Control, and Quality Assurance.

In approaching this task, an action within a support function should be selected. The action should be one that impacts the day-to-day work of production. A good example is a plant's order entry and production tracking process. This function is ideal for appling Value Stream Mapping, where each step of the process can be examined for waste and redundancy. If the mapping is completed properly, it's not unusual to cut the processing time in half or even more.

Many of the same techniques that are used to improve shop floor operations can be utilized in an Office Kaizen. The fundamentals of Workplace Organization are especially effective. The process normally involves a three-to-four day training session, along with the development of a plan of action, followed by a break of anywhere from one-to-three weeks to implement the action plan fully. Upon completion, a two-day session — a re-gathering of the participants —looks at the overall accomplishments and assigns follow-up responsibilities. At a minimum, one such event should be conducted in each of the plant's major supporting functions.

ACTIVELY INVOLVING SUPPLIERS IN THE PROCESS

Ideally, any number of key suppliers should be invited to participate in the training and implementation events conducted at the factory. Their thoughts and ideas can be extremely valuable, because they are typically extended without preconceived notions regarding the way things should be done — in other words, without blinders on.

Building a proper vendor relationship can take years. Efforts should be made early in the process to have a select number of suppliers deliver their goods without being required to go through receiving/inspection. To do this, they will need some form of special vendor certification. But the long-term goal should be to have as many suppliers as possible deliver directly to the shop floor, in the precise quantities required. Although achieving this goal will take considerable time and effort, it should be the ultimate objective.

In order to get to *Level Three* status, a plant is required to have a supplier certification program fully intact. Ideally, requiring suppliers to delivery in the quantities noted should come at no added cost. But even if this aspect wasn't included in the initial certification standards, it can always be added as time goes by. Nothing should be completely static with respect to a company's expectations of its suppliers, as long as those expectations are fair and reasonable.

Toyota changed the course of history with respect to supplier expectations. They did so in a give and take manner, with a long-term relationship in mind. The U.S. automotive industry picked up on the merits, but proceeded to place emphasis on achieving the very lowest price possible. This emphasis came back to haunt the automakers in numerous ways, including the overall quality of the cars they produced.

A proper supplier relationship takes a great deal of work and is best achieved through a solid commitment to seeing it through to completion, which could include the need for added resources. But to make Lean Manufacturing all it can be, supplier certification has to be addressed and carried out to its fullest.

BUILDING LEAN THINKING INTO PRODUCT DESIGN

Another element of achieving Level to its fullest is building Lean thinking into new and revised product designs.

This is best accomplished through a "design for manufacturing" document, working in close conjunction with the plant's product engineering group.

Product Engineering will need to provide the assurance it will do all it can to build Lean thinking into new and revised product designs. Manufacturing can help by providing an outline of parameters, covering such information as preferred notch and angle dimensions, hole sizes, attachment hardware, part widths and heights, ease of assembly specifications and so forth.

We often hear manufacturing complain about product design. Seldom do we see them take the initiative to see that a design-for-manufacturing guideline is prepared and put into effect. Suppose, however, the Product Engineering group has absolutely no interest in helping to see Lean accomplished. Most often this disinterest happens not because they are actually unwilling to help manufacturing, but because their plate is already full of what they perceive as higher priorities. When this occurs, Manufacturing can start the ball rolling by preparing and submitting recommendations for design. This is something that can't be entirely ignored by product engineering and can go further by building their awareness and attention on the matter. Design for manufacturing isn't just related to Lean. It should be done regardless. But Lean is a good place to get the ball rolling if it isn't already in effect.

The Hyena

The Lion is defined as the king of beasts and has the reputation of courageously facing any challenge in order to feed itself and protect its young. The Hyena, on the other hand, posseses the skill to be a hunter and, in harder times, can and does shift to the role. But for the most part, it preys on opportunity and finds it much more convenient to beg, borrow, or steal. When all is said and done, it's the Lion that gets the choice cut and the Hyena that picks up the leftovers.

Toyota is indeed a roaring lion and produces excellent products, with outstanding quality. However, Toyota's products aren't the least expensive on the market, by any means, which would seem to imply that price isn't the decisive factor. But that's far from correct because effective pricing boils down to what a customer perceives as a bargain. Most often, quality and price go hand in hand. Customers are generally willing to pay more for what they believe to be better quality, if the design and reliability are superior to the competition. But make the product a little too difficult to obtain and price becomes secondary in importance.

At some point, a company that makes its own products has to decide if marketing or manufacturing is its chief driver. Some would say both. But there's always going to be a tendency to place more emphasis on one or the other. If a company places a higher value on how it markets and distributes than on how it goes about manufacturing, a Lean initiative is always going to take a back seat to other priorities.

The simple truth is that a company that produces the products it sells is first and foremost a manufacturing-oriented firm. Manufacturing is where the best talent should be placed and where any benefit of doubt should be applied, when it comes to budget and the like. Toyota has been able to apply better design and reliability, better product quality, and better delivery principally because of the work it did and continues to do at a manufacturing level. I'm reasonably certain Toyota would agree with that observation.

All the best design, marketing, and distribution in the world simply cannot move a firm that is mediocre in manufacturing expertise to the forefront of a given industry. But moving on to greatness requires more than doing a decent job of implementing Lean. The company must do an outstanding job, which includes accepting the fact that all manufacturing operations can be improved. This acceptance is where *Level Four Lean* comes into play, and where a thoughtful continuous improvement process can be of extreme value.

PERFORMING A CORE PROCESS ANALYSIS

I addressed this topic in *Leading the Lean Initiative,* which centers on understanding which specific types of work a manufacturing firm should be doing or otherwise avoiding. No manufacturing firm can be great at everything. Therefore, the focus needs to be on doing the things that will set it apart from the competition.

Consider a firm whose key competitive edge rests in its woodworking and assembly skills. The plant also does its own plating and finishing, which is inclined to excessive scrap, rework, and obsolescence. Keeping adequately-trained operators is difficult and as time goes by, the firm's core processing starts to suffer as an ever-growing amount of capital, expense, and manpower are diverted to plating and finishing.

If someone decided to take a look at the actual costs, they would likely find the assigned product cost did not reflect actual expenses. If the investigation was carried further, there's a good chance of finding another firm in the vicinity that could do the work for less and with equal-to-better quality.

No good company would knowingly perform work that was clearly inferior to the competition. The key to performing a Core Process Analysis is to know what sets an operation apart and to identify areas where they are unquestionably better than the competition. Where the evidence indicates a firm's operations are _clearly_ better are where the principal focus should be directed because the idea is to *stay better*. But in areas where evidence indicates the firm is behind its competitors, serious consideration should be given to the drain placed on a company's core expertise. If they are substantial, which can sometimes be the case, then serious efforts should be directed at outsourcing the work, in a cost-effective manner.

Becoming Lean requires understanding what a firm is good at doing, then subsequently focusing efforts on being the best at everything it does. A Core Process Analysis can go a long way in achieving this goal. The analysis involves answering eight process-related questions. Evaluating the answers to these questions will provide a good understanding of where to place efforts for a formal make vs. buy study. The eight questions are:

1. Does the process create an excessive amount of downtime, scrap, or rework?
2. Does the process require a number of special skills and classifications?
3. Is it difficult to maintain the necessary talent required to perform the work?
4. Does the process dictate the need for a Kanban?
5. Is the process inclined toward costly upkeep and maintenance?
6. Do standard costs for the process represent actual expenditures?
7. Does in-house talent readily exist to take care of maintenance and upkeep or is external help required to keep the process running?
8. Does the process consistently perform to manufacturer specifications, in terms of quality, reliability, and throughput?

Excessive Downtime, Scrap or Rework

A process with a history of excessive downtime, scrap, or rework should be evaluated in the terms of the reasons why. If these problems are due principally because of the sheer nature of the equipment and the required expertise to keep it running effectively, the process could be a candidate for outsourcing.

Special Skills and Classification

If a process requires unique skills or the need for special labor classifications, it could be a candidate for outsourcing. This is especially true for a process that doesn't fit the category of core expertise.

Difficulty Maintaining the Talent Needed to Run the Process

If it is difficult to keep the talent required and a persistent problem exists with recruiting the needed expertise, the process could be a candidate for outsourcing.

Creation of a Kanban (Push) Requirement

If the process has to run in batch format due to various restrictions — such as lengthy cycle times or long setups which cannot be fully addressed and resolved with the application of SMED — the process could be a candidate for outsourcing.

Costly Upkeep and Maintenance

If a process requires higher-than-ordinary upkeep and maintenance and is prone to excessive downtime, it should be evaluated for outsourcing

Standard Cost Issues

If the standard cost for a process fails to represents what the actual costs are, the process should be considered as a candidate for outsourcing. What establishes misrepresentation is largely up to an individual firm. Unless an operation wants to continue to experience unfavorable labor variance, it has to either adjust its standard cost accordingly, and thus raise the selling price of the product, or find someone to perform the work for the price spelled out at standard.

External Expertise and Equipment Specifications

If a process requires a great deal of on-going external expertise in order to keep equipment functioning properly, it should be marked for evaluation. This would not apply to recently purchased equipment that is still in the process of being debugged. However, questions should be raised as to why such equipment wasn't properly brought up to acceptable specifications before it was shipped to the factory.

THE PROPER ORGANIZATION FOR LEAN MANUFACTURING

Examining the organization and adjusting accordingly is extremely important to a good Lean Manufacturing effort. This goes for the operation as a whole, but the first and most critical position is always the plant manager. This position typically holds the responsibility for the factory in terms of output and site profitability. Experience has shown there are three basic tiers of plant managers; each carries varying degrees of impact on the process.

Tier One: Plant managers who are ultimately in charge of the entire factory and are clearly viewed as the site manager.

Tier Two: Plant managers who report to a person in charge of the site, who hold them responsible for meeting a prescribed schedule, which they may or may not play a major role in establishing.

Tier Three: Plant managers who report to a second tier within the site organization; who are, in essence, little more than production superintendents.

For factories that have Tier One plant managers, it is essential that these individuals are strong Lean advocates. Tier One plant managers are fully in charge of the factory. Once a Lean effort has been started, they become the catalyst for its overall success or failure.

For Tier Two and Tier Three plant managers, the power and influence of the position carries far less impact. If the person they report to is a Lean enthusiast, these plant managers will usually go along with the effort, whether they truly believe in it or not. If their boss isn't a strong advocate of Lean, such managers face an uphill struggle implementing the process. In most cases, success comes down to personalities and the freedom given, or otherwise assumed, to make changes in the plant's operations.

Once a go is given to a Lean Initiative, serious thought has to be directed at the matter of reporting links and the influence these links can have on the process. This topic was discussed in Chapter Four, under *Getting Middle Management on Board*. An operation cannot afford to have a person of influence who undermines the effort, whether openly or discretely. I've often told clients they need to approach this important matter as if the factory was drawing its last breath and Lean was the only solution. The past performance, loyalty, and accomplishments of a given individual must be carefully weighed against the need to get the job done in the most effective manner.

THE IMPORTANCE OF CHANGE

Knute Rockne was a legendary head football coach at Notre Dame. Great would be less than an adequate description of his coaching and the "Notre Dame Shift" he personally devised. In the shift, all four running backs were in motion at the snap of the ball. Opponents were so dumbfounded, they couldn't find a way to defend against it. Among the most noteworthy accomplishments about Rockne's 13-year tenure at Notre Dame were five unbeaten and untied seasons, twenty first-team All-Americans, and a lifetime winning percentage of .881 — including winning the last 19 games he coached.

With the Shift, Rockne knew he was on to something. Convincing him to change would have been difficult, if not impossible. But that same technique could not be used on a football field day today. The reason is that the rules have changed and overall football strategy has been altered dramatically.

One can look at the world of manufacturing in much the same way. What once proved to be a winner, given the timeframe, can today be a totally obsolete process. Assume the market share of a company that isn't into Lean *has not* been on the decline, that operating costs *have not* risen significantly, and that customer's *appear* to be reasonably satisfied. How does a company come to accept the need for change in the way it goes about implementing Lean? The most important reason is to gain the substantial competitive benefits associated with taking Lean Manufacturing to its ultimate level of accomplishment.

Think about the benefits associated with having your key production equipment fully engineered in SMED, Poke-Yoke, and TPM, and how much further these improvements would serve to take you in effectively completing the journey. It will ultimately come down to how strongly leaders feel about the need to shift priorities and how much faith they have in the approach being taken. I can assure anyone, without the slightest misgiving, that time isn't on our side.

A CLOSING COMMENT REGARDING THE TASK OF IMPLEMENTATION

If you closely examine the typical approach used to implement Lean, you will find the absence of a needed framework. This absence explains why some firms throw their efforts into implementing *a specific tool*, such as Kanban, rather than striving to insure the fullest possible incorporation *of the process*, across the entire factory. It is also why many firms seriously falter and find themselves with little to show for their effort a year or two down the road.

Unfortunately, far too many Lean undertakings are left entirely to chance and preoccupation. The undertaking becomes chance when a clear and successful path of accomplishment has not been constructed. It becomes preoccupation when the leaders of the process allow a certain field of endeavor to become a roadblock to implementing Lean Manufacturing to its fullest.

Setting sight on fully achieving Level One Lean before moving on is a measurable and effective means of accomplishing the task. Making the first step requires a focused dedication to engineering a plant's key production equipment. In the course of doing so, implementation can proceed in a much smoother and more effective manner.

When it's all said and done, assuring the long-term success of the U.S. manufacturing base will depend largely on making Lean a full and absolute success, across a broad span of industry. This book has pointed to a means of doing this, through an approach aimed at properly setting the stage, adequately measuring progress, and charting a clear course for implementation.

I do not expect what has been outlined to suddenly become the preferred and accepted means of implementing Lean because we're speaking about a substantial change in approach. However, if this book does nothing more than inspire a higher awareness for a more structured method of implementation, much of the goal I set out to achieve will be accomplished. It's my sincere hope that a seed has been planted which will take root and play a role in bringing manufacturing in the United States back to the prominence it once enjoyed — for as broad as the challenge we are facing may be, where there's a will, there's a way.

KEY REFLECTIONS

- Toyota read Henry Ford's book, listened to Juran and others, and came to the United States to observe our processes for themselves. With that knowledge in hand, the springboard for the Toyota Production System was born.

- This, in turn, created a framework for what was generally expected of the hourly worker. Keep your nose clean, show up for work on time, check your brains at the door, and do your job. Sound familiar?

- One of the more significant steps in achieving Level One Lean involves developing an 18—24 Month Implementation Plan and seeing it through to completion.

- Ideally, key suppliers would be invited to participate in the training and implementation events conducted at the factory. Their thoughts and ideas can be extremely valuable because they are typically delivered without blinders on.

- Toyota's products aren't the least expensive on the market, by any means, which would seem to imply price isn't a factor. However, effective pricing is based on what a customer perceives as a bargain. Most often, quality and price go hand in hand.

- Once support is given to a Lean Initiative, serious thought must be directed toward evaluating the existing reporting links and the influence they can have on the process.

- The undertaking's success becomes a matter of chance when a clear and successful path of accomplishment has not been constructed, and it becomes preoccupation, when the leaders of the process allow a certain field of endeavor to become a roadblock to implementing Lean Manufacturing to its fullest.

REFERENCES

- Schonberger, Richard J. *Building a Chain of Customers*. The Free Press, 1990.

- Abbeglen, James C. and George Stalk, Jr. *Kaisha: The Japanese Corporation*. Basic Books, Inc., 1985.

- Maccoby, Michael. *The Leader — A New Face For American Management*. Simon & Schuster, 1981.

- Wexley, Kenneth N. and Gary A. Yukl. *Organizational Behavior and Personnel Psychology*. Richard D. Irwin, 1977.

- Ohno, Taiichi. *Toyota Production System*. Productivity Press, 1987.

- Steinbacher, Herbert R. and Norman L. Steinbacher. *TPM for America — What It Is and Why You Need It*. Productivity Press, 1993.

- Davis, John W. *Leading the Lean Initiative*. Productivity Press, 2001.

- The University of Notre Dame. *Notre Dame Football Memories: Knute Rockne*. The Official Athletic Site of the University of Notre Dame, 2008. http://und.cstv.com/trads/rockne.html

INDEX